Small Farm Development

IADS DEVELOPMENT-ORIENTED
LITERATURE SERIES
Steven A. Breth, series editor

Small Farm Development
was prepared under the auspices of
the International Agricultural Development Service
with partial support from
the U.S. Agency for International Development

ALSO IN THIS SERIES

*Rice in the Tropics: A Guide to the Development
of National Programs,* Robert F. Chandler, Jr.

Small Farm Development
Understanding and Improving
Farming Systems in the Humid Tropics

Richard R. Harwood

Westview Press / Boulder, Colorado

Cover photo by Ray Borton

Copyright © 1979 by the International Agricultural Development Service

Published in 1979 in the United States of America by
  Westview Press, Inc.
  5500 Central Avenue
  Boulder, Colorado 80301
  Frederick A. Praeger, Publisher

Library of Congress Cataloging in Publication Data
Harwood, Richard R.
  Small farm development.
  (IADS development-oriented literature series)
  Bibliography: p.
  1. Farms, Small. 2. Agriculture—Tropics.
  I. Title. II. Series: United States.
International Agricultural Development Service.
IADS development-oriented literature series.
HD1476.A3H36   338.1'0913   79-13169
ISBN 0-89158-669-5
ISBN 0-89158-699-7 pbk.

Printed and bound in the United States of America

Contents

List of Figures and Tables ix
Foreword, *A. Colin McClung* xi
Preface xiii

Part 1
Small Farm Development

1. A New Approach to Analysis 3

   A contrast in development approaches 3
   An effective small farm development method 6

2. The Stages of Small Farm Development 9

   Stage I: Primitive hunting-gathering 11
   Stage II: Subsistence-level crop and
     animal husbandry 11
   Stage III: Early consumer 12
   Stage IV: Primary mechanization 15
   Development effects 16
   Development within or across growth stages 17

3. Goals of Small Farm Development 21

   The uses of labor 22
   Long-term versus short-term goals 23
   Cultural aspirations 25
   Appraising development goals 26

4. Measuring the Well-Being of Small Farmers 27

  Need for indicators. 27
  Cash flow as an indicator 29
  A comprehensive index of well-being 29

5. Research in Small Farm Development 32

  Experiment station focus 32
  Farmer focus 33
  A philosophy of collaborative research 34
  On-farm research 38

Part 2
Critical Factors in Small Farm Development

6. Physical Limits to Cropping Intensity. 45

  An example: Water 46
  Cropping pattern potential and water availability . . . 50
  Importance of topography to water conditions 56
  Temperature 58
  Tillage capability 58
  Soil fertility 59
  Environmental classification systems 60

7. Economic Determinants of Crop Type and
      Cropping Intensity. 63

  Labor 64
  Management capability 66
  Power 67
  Cash 69
  Market availability and subsistence production 70
  Secondary factors 71

8. Resource Requirements of Multiple Cropping 76

  Crop sequencing. 76
  Relay planting 83
  Intercropping 85
  Perennial crops 90

9. Animals in Mixed Farming Systems 93

   Contribution of animals to mixed systems 93
   Sources of feed 94
   Animal management 97
   Advantages of mixed systems....................99

10. Noncommercial Farm Enterprises 101

   The farmyard as a center of production 101
   Fencerows.......................................103

11. Nutrient Needs of Intensive Cropping Systems......106

   Purchased fertilizers............................106
   Nutrients collected from outside the farm 107
   Recycled farm materials 107
   Nutrients recycled within each crop..............109

12. Efficient Use of Farm Resources..................115

   Farmer priorities for resource use................115
   Fertilizer 116
   Farm power 116
   Crop diversity and management 117

13. Requirements for Farm Mechanization 119

   Primary mechanization..........................119
   Secondary mechanization 121
   Transportation 123
   Power and farm resource use.....................123

14. Stability in Farming Systems......................125

   Biological stability..............................126
   Management stability 127
   Production stability 129
   Economic stability 129

Appendixes

A. Sources of Farming Systems Information 133

B.   Farming Systems Terminologies 136
C.   Botanical Names of Crops Mentioned 141

Annotated Bibliography.............................145
Index 155

Figures and Tables

Figure

1. Labor productivity, number of farm
   enterprises, cash investment, and skills
   required in different agricultural growth
   stages when markets for high-value crops
   are limited 10
2. Conceptual model of the production system
   of a Nepalese hill farm 18
3. Categories in a rainfall classification for
   rice-growing areas of Asia. 48
4. Water availability and field crop patterns for
   upland rice farms of eastern Batangas,
   Philippines 51
5. Alternative cropping patterns for eastern
   Batangas, Philippines 52
6. Possible crop patterns in lowland rice areas
   having three to five months of good rainfall 55
7. Physiographic classification of rice paddies 57
8. Mung bean responds to low rainfall: yield in
   relation to rainfall between planting and
   first harvest 62
9. Relationship between farm size and type of
   crop in areas of Southeast Asia having a
   two-crop (eight-month) growing season
   and access to markets for high-value
   vegetables 72

10. Change in total farm productivity through
    addition of various power sources in area
    where the growing season is six to eight
    months 74
11. Schematic relationship between farm size and
    power source commonly found on developed
    farms in Southeast Asia 75
12. Cash returns to labor in relation to method of
    meeting crop nutrient requirements.............110
13. Nutrient cycling in cropping systems 111
14. Power-labor ratios needed for various crop
    enterprises 117

Table

1. Characteristics of development stages in
   agriculture (for farms with a relatively high
   level of resource use for their development
   stage)..19
2. Types of cropping patterns commonly used
   in Asia, their energy requirements and
   productivity characteristics....................92

Foreword

In much of the tropical world, when one looks over the countryside, he sees not uniform fields of waving grain but a patchwork of small fields containing mixtures of crops. And even in regions where a crop like wheat or rice dominates the landscape for a few months, farmers are likely, immediately after harvest, to plant a totally different crop or combination of crops. The small farmer in the tropics employs intricate farming systems to adjust to seasonal changes in temperature, rainfall, marketing conditions, and the availability of family labor. Through these systems, also, he survives the unpredictability of his environment.

The interactions in tropical farming systems are complex. A small change made at one point in the system may set off far-reaching tremors elsewhere in the system. Science has much to contribute to these farming systems. But to do so, researchers must be unusually adept at seeing the world from the farmer's vantage point.

This book is intended, as the subtitle suggests, to help agriculturalists and others understand farming systems in the humid tropics so that those systems can be improved. The world's resources of uncultivated land are dwindling. Food for future generations will come largely from making already-cultivated land more productive. Higher productivity will result not only from varieties that have a higher yield potential, but from capturing, in economic crops, more of the sunlight and water that strike the land, and by fostering the subtle natural interactions between animals and crops, and between crops themselves, that favor higher and more stable yields.

Richard Harwood has long experience with the cropping patterns of small-scale farms in the tropics, as a crops researcher for the Rockefeller Foundation in Thailand, as head of the multiple-cropping project at the International Rice Research Institute, and as a consultant on tropical farming systems.

This book is part of the IADS Development-oriented Literature Series. The aim of the series is to bring together, concisely, authoritatively, and readably, up-to-date information related to agriculture in the tropics and subtropics for policymakers, advisors, and others.

The development of this book was financed by a grant from the U.S. Agency for International Development.

A. Colin McClung
Executive Officer, IADS

Preface

In our impatience with "backward" small farmers and in our haste to rapidly "commercialize" them, we have overlooked key aspects of their farming systems that could enhance our efforts to increase food production and improve rural well-being. To accomplish the development of a greater number of the world's small farms, shifts in emphasis must be made in our thinking, in our technological research, and in our communications with farmers. The central theme of this book involves analysis of several aspects of small farm production systems that increase efficiency when the farmer's production resources are limited. The purposeful blending of traditional and modern technologies may well prove the key to starting the most disadvantaged farmers along a more rapid development path.

Better understanding of small farm systems will encourage rational investments and infrastructural changes that will result in more effective development efforts. This book is conceptual in nature and is meant to influence development thinking rather than to provide a detailed how-to-do-it guide.

The five chapters of Part 1 present an overview of small farms and the options for their change. Chapter 5, the heart of the book, suggests a development approach for collaborative work among scientists, extension workers, and farmers to both develop and extend relevant technology to resource-limited farms. Chapters 1 through 4 summarize small farm types and their production systems. In Part 2, Chapters 6 through 14 deal in more depth with critical aspects of small farm development that are either overlooked or given little emphasis in development programs. The appendixes give sources of additional information and define terms. Selected references are provided in the annotated bibliography.

The book is heavily influenced by my experience in tropical Asia, but relevant examples from other parts of the humid tropics are included, and the principles described are universal in applicability. The discussion applies most directly to those areas of tropics having rainfall greater than 1000 millimeters per year.

I am indebted to the many colleagues and small farmers of the developing world who have contributed so greatly to my firsthand experience with small farm agriculture over the past 10 years. My involvement has been people-centered, to the enrichment of my own life. Recognition is given to the programs of the International Rice Research Institute and to the International Development Research Centre whose funding and interest made possible much of the work referred to here. The Rockefeller Foundation and the International Agricultural Development Service were exemplary in making it possible for me to write this book.

To Dr. Raymond Borton I owe thanks for many hours of editorial work and critical comment. His contribution of the cover photo and his considerable technical input to Chapters 1 through 5 are gratefully acknowledged. Anthony Wolff did the final editing.

Richard R. Harwood
Emmaus, Pennsylvania

Part 1

Small Farm Development

A New Approach to Analysis

The factors that limit food production on the world's small farms are virtually unlimited in number and variety. The small farmer does not have enough land to produce more; or family labor is scarce; or his family feels a competing need for cash income from nonfarming pursuits: the possibilities are endless. Land is the first limiting factor in most areas: more than 90 percent of all tropical farms are less than 5 hectares in size. National averages in Asia are often less than 3 hectares, as in the Philippines; or between 1 and 2 hectares, as in Bangladesh. Low soil fertility and poor soil structure, poor seed, water shortages, extreme temperatures, lack of access to inputs and markets—all limit the capacity of the small farmer in the tropics and subtropics to produce food.

A contrast in development approaches

Faced with these multiple limitations, each a formidable problem, agricultural development programs inevitably tend to concentrate their efforts on those few factors that seem most crucial to crop production and easiest to improve. The resulting advances—the development of high-yielding varieties of key grain crops, the proliferation of irrigation systems, and the widespread introduction of fertilizer and other inputs—have helped greatly to keep national food production in the developing countries more or less in step with rapidly rising demand. So far, however, these production increases have come largely from the most favored farming areas, where

In Nepal farmers use their scarce land intensively. The farmstead areas have carefully tended vegetable gardens and fruit trees. Trees bordering cultivated fields provide compost, animal fodder, and firewood.

the constraints on production are relatively light. However, the continuing need for more food production and the growing concern for the well-being of the small farmers who have been largely untouched by the new technologies are drawing attention to the special problems of small farmers in the tropical and subtropical developing countries.

When resources are limited, the key to farm productivity, and thus to the well-being of farm families, is the interaction of varied but complementary farm enterprises. Analyses of these interactions, however, have traditionally focused on larger farms and emphasized labor productivity and return on investment as critical variables. The small farmer in the tropics seldom enjoys the option of varying his capital.

Also, traditional development programs have often been aimed at a single commodity. Not surprisingly, they have been most successful in situations where farmers depend predominantly on a single food grain, and where there is a profitable market for their production. The small farmer often finds such programs irrelevant or unacceptable because they do not encompass the varied mix of crops and livestock that is his daily concern, and because they put him at the mercy of market forces he cannot control and probably does not understand.

This brings us to a distinction between farm development as proposed in this book and its common use in today's development programs. Farm development is usually considered synonymous with commercialization. The most frequently stated objective of today's programs is increased farm income. Other indicators of development progress are amounts of cash inputs used and farmer participation in credit programs. The underlying assumption is that greater cash flow across the farm boundaries (increased commercialization) is a true indicator of increased farm productivity and improved farm family well-being.

Our slowness or outright inability to commercialize large segments of the world's farmers and the questionable effects of such commmercialization on family well-being in other cases lead us to a more general concept of development for small farms. Farm development as used here signifies a progression to more efficient and more productive use of limited farm resources. It nearly always implies an increase in labor productivity and an increase in quality or quantity of the food and fiber output of a farm unit. In the early growth stage, in particular, it probably will not involve commercialization.

In contrast to traditional approaches, the analysis and attack proposed in this book are based on the agricultural systems actually used by small farmers in tropical areas. The farming system is a set of biological processes and management activities organized with the available resources to produce plant and animal products. The farmer's resources include such physical factors as soil, sunlight, and water, plus such economic and social factors as cash and credit, labor, power, and markets. The limits of the analysis in this book are strict: the farmer himself and the resources he has to work with on his small land area. Accordingly, the analysis includes marketing activities only to the point where the product reaches the first off-farm handler. Processing activities are included only when the crop requires preparation for first-stage marketing, as in the case of tobacco, which must be dried, or grain, which must be threshed.

An effective small farm development method

The analytic process described in the following chapters is evolving constantly; it must be adapted to each agricultural environment and farming system to which it is applied. It is not an ideal, mathematical system of precise measurement and by-the-book interpretation. Such an academic system would necessarily be irrelevant to the small farmer's situation.

The analysis of farming systems properly begins with the identification of significant interactions: of people with plants, plants with animals, plants with other plants, and so on. Each interaction propagates others; the problem for the analyst is to sort out the various reaction products, define the significant ones, and match them with the farmer's goals. To do this, the analyst must order the variety and complexity that are characteristic of tropical farming systems. The aim of the process is to identify situations in which existing farm resources are inefficiently used. The process succeeds when it defines changes in the farming system that result in increased productivity.

To understand the farmer's agricultural system, the analyst must classify the various environmental factors to which the farmer responds and identify local farms where these factors are expressed in varying degrees. The basic factors to be analyzed are soil and climate. Local crop and animal production data are also important. These factors must be viewed in terms of the farmer's own goals and priorities, which figure as importantly as objective physical and biological considerations in his decisions about how to farm. His need for food, his competing need for cash income, his status in the community, his desire for stability and security, his motivation to conserve energy and other resources—such subjectively perceived values are also factored into the farmer's agricultural equations.

Only after he has gathered this information and understood its meaning for the farmer can the analyst plan appropriate changes in the farming system. The planning process involves the scientist with the farmer in deciding what modifications and innovations to try. Each brings to the planning process his own perspective and his own wisdom. The farmer contributes his intimate, often tacit, understanding of his own situation and the factors that influence his productivity. The scientist has the objective information derived from his measurements and observations, plus a familiarity with alternative production technologies from other areas. The scientist and the farmer collaborate on planning and implementing changes, and the results are measured against mutually agreed-upon goals. The careful documentation of their experience with new technologies and systems in well-defined environments makes it possible to extrapolate their results to other, similar situations in any part of the world.

This approach depends to a great extent on teamwork among scientists whose disciplines are highly specialized and insular and who are unaccustomed to working together on common problems. The process proposed in this book requires agronomists to work with crop and soils scientists, animal specialists, agricultural economists, nutritionists, and educators. Interdisciplinary collaboration is crucial to the process, and the team includes a coordinator whose special function is to bring the disparate insights and skills of the various scientific specialists into focus on the problem of increasing the small farmer's production.

2
The Stages of
Small Farm Development

Applied to a farmer or a group of farmers, common adjectives like "small," "bypassed," "underdeveloped," or "disadvantaged" do not convey much useful information. Indeed, they obscure more than they reveal; critical distinctions are lost. Such euphemisms fail to specify the essential differences between the farmers they purport to describe and their more favored cousins; nor do they indicate the diversity and complexity that characterize small farm agriculture.

The descriptive classification of farming systems used in this book is based on stages or levels of development, according to the system's physical environment, local food habits, the availability of inputs and markets, and other factors (Fig. 1). Each of the developmental stages in this classification can be characterized by any of several combinations of crops and livestock, and the examples given are meant to be illustrative rather than exhaustive.

A descriptive taxonomy such as this one recognizes that agricultural development proceeds through a definite series of growth stages, though not always at a steady rate or even continuously. The pattern of development changes according to whether the basic production is cultivated crops, tree crops, or livestock. Cultivated crops provide the bulk of farm products, and they will receive the most detailed treatment in this analysis; but in the tropics a typical farm of 3 to 5 hectares includes a combination of cultivated crops, animals, and tree crops.

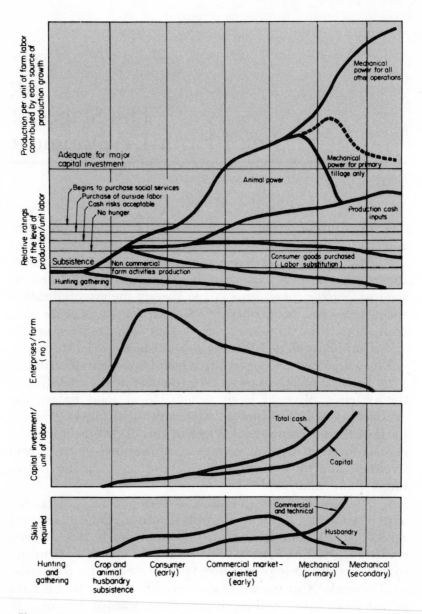

Figure 1.
Labor productivity, number of farm enterprises, cash investment, and skills required in different agricultural growth stages when markets for high-value crops are limited. (Source: IRRI)

Stage I: Primitive hunting-gathering

Until the recent discovery of the Tasaday tribe of Mindanao in the Philippines, hunting-gathering as the sole means of subsistence was believed to be extinct. Partial hunting-gathering systems are still found in many societies, however. Indeed, most farmers who are remote from markets rely on hunting and gathering for some part of their total food production. Throughout Southeast Asia, for example, hill tribes such as the Rhade of Vietnam combine crop and livestock cultivation with the harvest of roots and fruit from the forest. The Rhade diet also includes an occasional wild animal, as well as insects, frogs, crayfish, and small rodents.

Hunting-gathering as an element in a more advanced farming system is also found among Nepalese hill farmers, who gather leaves from nearby forests to compost with animal manure into fertilizer for their intensively farmed land. The labor required by such hunting-gathering activities is varied enough to employ family members of all ages, including children and older people who might otherwise be unoccupied.

Despite its apparent simplicity, hunting-gathering can destroy the natural resource base when population pressure exceeds the ability of the environment to renew itself. In parts of India, for example, all the leaves are stripped from the rapidly declining stock of live trees to provide fodder for the burgeoning population of sheep and goats. When hunting-gathering reaches this stage of resource exhaustion, the productivity of the labor it requires is very low, especially when measured in terms of the cash value or opportunity cost of the time invested in each unit of production.

Stage II: Subsistence-level crop and animal husbandry

Subsistence farming is still common today in remote areas. At this level, more than 90 percent of the farm production is consumed directly on the farm; there is little selling or trading. Such noncommercial farming systems are excluded from any development process or program that involves cash income, marketing of farm products, or purchase of inputs.

Among the subsistence farms of Asia, the number of shifting cultivation farms (which cover roughly 40 percent of Asia's total crop area) is about equal to the number of fixed or permanent farms. In shifting systems, land area per farm is greater, intensity of land use is lower, and there are fewer crop-animal interactions than in fixed subsistence systems. Fixed subsistence systems, on the other hand, have adapted to higher population pressure on land and have increased resource-use efficiency through more and more "structure" in the farm system.

Subsistence farmers typically produce a great variety of crops and animals. In Asian monsoon areas with more than 1500 mm of annual rainfall, it is not unusual to find as many as 20 or 30 tree crops, 30 or 40 annual crops, and 5 or 6 animal species on a single farm. One group of subsistence farmers on Mindoro Island in the Philippines regularly depends for food on a total of 430 plant species. By the same token, subsistence farming is characterized by diverse labor requirements. Having evolved to produce food year-round, the system provides continuous employment for unskilled labor to tend crops and livestock.

Typically, the subsistence farmer plants some crops—rice, for example—that are preferred by the community but that entail relatively high risks. He hedges against this risk by growing several less valued but also less uncertain crops, such as cassava. In monsoon climates, with pronounced alternations of wet and dry seasons, the subsistence farmer ensures a stable production with long-duration root crops, tree crops, and animals. While it lacks the potential for producing a marketable surplus, and thus supporting a higher standard of living for the farm family, subsistence farming has real strengths. The crop and animal combinations evolved by subsistence farmers can often be adapted to increase productivity on more highly developed farms that are pushing against resource limitations.

Stage III: Early consumer

At the early consumer stage of development, the farmer mar-

A newly "cleared" slash-and-burn field containing a mixture of taro, cassava, and young fruit trees. The productivity and stability of this system depends on how soon after clearing the farmer succeeds in planting perennial crops.

kets between 10 and 30 percent of his production, and the resulting cash income enables the farm family to avail itself of goods and services beyond the barest necessities. Besides salt and oil for cooking or lamps, the family may buy cloth instead of weaving its own. It may also buy porcelain dishes instead of using homemade pottery. The income from surplus food production may also be reinvested in iron farm implements and other capital improvements. C. R. Wharton has estimated that 60 percent of the world's farmers market less than half of what they produce.

The consumption of the farm family at this stage is paid for by the productivity of family labor. Productivity is increased by several means: introducing tree crops, such as coconuts or coffee, that yield well with relatively little labor; planting valuable market crops, such as tobacco or vegetables; and using animal power to speed tillage, transport, and harvesting. The change in development stages is exemplified by the shift from upland rice to lowland paddy rice, in which a surplus crop of marketable quality is produced by the intensified use of human and animal power and by improvements in irrigation and tillage.

To become a consumer, even at this low level, a farm family must first rise above the hunger level and be able to accumulate a production surplus that can be turned into cash income. The family gradually increases its crop and animal husbandry skills and usually diversifies into a variety of enterprises. The increasing need for cash income becomes a powerful motive for the family to concentrate its labor and inputs on market crops. There is a progressive shift of labor toward more productive enterprises, even though actual cash outlays may still be quite small. The labor that is no longer devoted to weaving crude cloth or making clay pottery at home is diverted to surplus farm production for the market. These changes occur even as elements of hunting-gathering and subsistence farming persist.

The early consumer stage begins in earnest when farmers invest the cash income from the sale of their surplus production in inputs to increase production even further. In general, farmers hesitate to risk such cash investments until certain basic conditions exist:

- The farmer has adequate food of acceptable quality for his own family.
- He has confidence in his own technical, agricultural, and commercial skills.
- His farming system is ready to respond to additional inputs.
- He has the technology to turn inputs into increased production.
- He has access to markets and to the cash economy.

In short, a farmer must enjoy a certain level of well-being

before he will commit himself to the early consumer stage of development. This threshold level varies somewhat among cultures and areas, and its determination is one of the basic issues in the analysis of any farming system. On Siargao Island in the Philippines, for example, studies revealed that improvement in rice culture was contingent on the food intake of the farm family. Without adequate food, there was no extra margin of human energy for weeding, improved land preparation, and transplanting. Moreover, although the island's farmers knew about fertilizer and its advantages, and presumably would have used it, they had no surplus production from their subsistence farming to sell for cash to purchase inputs.

The potential of all critical elements in the farming system to respond to extra effort and inputs is a prerequisite for progress to the consumer stage of development and beyond. If water is a limiting factor, no investment in fertilizer or pesticides is likely to be profitable. Improved plant varieties must be adapted to the particular conditions of the area. Water supplies and climate conditions must be reliable to minimize risk. Each of these elements is a vital component of the total farming system, and each must be accorded its actual weight in any relevant analysis.

It is often suggested that lack of credit with which to purchase inputs at the appropriate time is a major factor limiting small farm productivity. We are suggesting here, however, that farm development must already have progressed to a certain point before credit can be useful in increasing production. The availability of power—animal or mechanical—is another precondition for development, though its importance is sometimes overrated. Without power, the farmer is limited in the kinds of enterprises that will reward his labor enough to be worthwhile, unless he has a market for high-value specialty crops. Lacking adequate power, the farmer may turn to low-labor crops such as coconuts, oil palms, or rubber.

Stage IV: Primary mechanization

The farmer has reached the primary mechanization stage

when he rents or purchases a source of mechanical power. Mechanization and commercial farming proceed hand in hand, accompanied by a number of parallel changes. The number and diversity of enterprises on a single farm declines, after proliferating in the previous stages. With this simplification comes a decline in the number and variety of husbandry skills needed by the farm family. At the same time, however, the farmer's need for commercial abilities and technical skills increases sharply. Noncommercial farm production declines as the available capital and labor are increasingly invested in commercial crops.

Where labor scarcity limits production, the introduction of mechanical power to amplify human effort may be the key to further development. In some cases, mechanization can free labor for off-farm employment that can provide both the stimulus to increase labor productivity and the cash income to buy or rent the necessary mechanical power.

The recent worldwide increases in energy prices have prompted concerns about the prudence and propriety of increasing dependence on farm machinery as a substitute for human or animal power. The actual energy costs per unit of production, measured in calories, are approximately equal for human labor, animal power, and mechanical power. In view of the increasing scarcity and cost of fossil fuels, therefore, it may be that animal power, rather than machinery, is the wisest choice for the immediate future on many of the world's small farms.

Machinery on small farms is usually used first for transportation and then for primary tillage. Irrigation and threshing are also early candidates for mechanization. Mechanical cultivation, planting, harvesting, and processing follow in a secondary mechanization stage.

Development effects

From the foregoing brief description of the progression of development stages it is clear that advances in development are closely correlated with increases in labor productivity (Fig. 1). Central though it is, however, labor productivity cannot be used as an index to the general well-being of farmers without reference to other important factors. A farmer whose labor productivity is relatively high, but who purchases all of his family's food, especially in some remote areas where food costs are high, may have a far lower living standard than a less productive farmer who raises all or most of his family's food. The typical Southeast Asian smallholder who grows rubber, for example, enjoys relatively high productivity, but suffers from high living costs and a relatively low net income.

Development within or across growth stages

A major hypothesis of this book is that development can occur within a growth stage as well as across growth stages. The common assumption that development is synonymous with commercialization is the very error that leads us to ignore the great majority of poor farmers. Considerable development can, and often must, occur before the commercialization process begins.

The hill farmers of eastern Nepal, for example, are extremely isolated. They have little or no market available, and no purchasable production inputs. The short growing season and large extended families strain the productive resources of the small farms. These noncommercial production systems are highly organized interactions of crops with other crops and crops with animals, with the interactions being crucial to their productive efficiency. Cropping intensity has increased to the point at which soil fertility becomes the primary limiting factor. The animal component of the system is critical to its productivity. Animals contribute to nutrient cycling as well as to farm power and the family food supply (see Fig. 2).

Because of the extreme isolation of the farms (as much as 1000 km from a road) it is difficult to alter the basic components of the system, but better management of forest and grazing lands and introduction of improved varieties could add to production. Organization of markets for surplus and using cash earned from selling a surplus for fertilizer to supplement the compost (limited commercialization) would be the next step. But in many parts of Nepal where roads have been built and fertilizer made available farmers have been reluctant to use cash inputs. Perhaps the progression from complete subsistence to

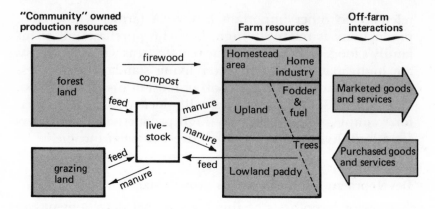

Figure 2.
Conceptual model of the production system of a Nepalese hill farm.

commercial farming requires a gradual transition through the consumer stage with an increase in cash flow for consumer goods before the farmer is willing to invest in a cash production input. Until the system generates a cash surplus, the availability of roads and inputs does not in itself bring development to such an area.

By contrast, subsistence farmers in remote areas of much of Southeast Asia have far fewer husbandry skills than the Nepalese farmers just described. Shifting cultivation is common, with its resource-exploitative practices and lack of biological balance. Animals are left to wander with little management. Symbiotic animal-crop support relationships are nearly absent. In fact the wandering, untended animals prevent the growing of many vegetable and fruit crops around the home. Despite having far more favorable land and climate, the farmer in remote areas of Southeast Asia has a much less stable and less productive system than his counterpart in remote areas of Nepal. The shifting system, because of its lack of structure and lack of component interaction, makes far less efficient use of limited resources than does the fixed subsistence system of the Nepalese farmer. Because of low labor productivity, the shifting farmer also has less time to spend on housing improvements which increase his well-being.

Table 1 illustrates the change in the structure of the

Table 1.
Characteristics of development stages in agriculture (for farms with a relatively high level of resource use for their development stage)

	Shifting cultivation	Permanent agriculture (subsistence)		Commercial family farms	Corporate or state farms
		Less than 10% sales	10-50% sales	over 50% sales	
Proportion of farmers involved	over 40%			less than 50%	less than 3%
Predominant labor activities					
Landclearing	x				
Tillage by hand	x	x	x		
Tillage by animal		x	x	x	
Tillage by machine				x	x
Animal tending		x	x	x	
Crop tending	x	x	x	x	x
Nutrient cycling		x	x		
Harvesting	x	x	x	x	x
Marketing			x	x	
Types of farming systems					
Monoculture crops	no	yes	yes	yes	yes
Intercropping	yes	yes	yes	rarely	no
Draft animals	none	yes	yes	yes	none*
Pigs untended	yes	no	no	no	none*
Poultry untended	yes	yes	yes	yes	none*
Complementarity of interactions between crops and between crops and animals	slight**	very high	high	moderate	slight
Importance of farmstead to family nutrition	slight	very high	high	moderate	slight

*Animals and cultivated crops are usually not mixed on corporate farms in the tropics.

**Negative when animals compete with people for food.

farming system that occurs as shifting agriculture develops through the permanent subsistence stage and on to the commercial stage where resources are limited. With more intensive land use, interaction between farm enterprises increases. Crop-animal interactions reach a maximum in the noncommercial permanent systems. With commercialization these interactions decrease in importance.

Most of the foregoing discussion has involved the obvious changes across development stages prior to commercialization for those farms that have reached a high level of resource use within their development stage. There are obvious and equally great differences in resource use and in productivity within a development stage. Few permanent subsistence farmers, for instance, have attained the higher level of well-being found on the better farms. Possibilities of growth within the early development stages are greater than we have commonly been led to believe. Improvement in rural well-being is possible before the massive infrastructure development required by commercialization becomes a reality.

Goals of Small Farm Development

The analysis of farming systems proposed in this book depends on identifying the farmer's goals as a necessary step toward devising alternative ways of reaching them. These goals are complex, varied, and usually tacit, however, and there is anything but unanimity among development specialists—or among small farmers themselves—on what the goals of development are, what they ought to be, or how they are to be attained. The basic agronomic, plant, and animal sciences on which agricultural development most heavily depends do not define the goals that development should pursue.

The fundamental goal of assuring enough food for the community and for the individual family is common to farmers in all rural societies. Beyond meeting basic food needs, the goals of families and societies become individual and diverse. This book is concerned with the goals of the individual farmer and his immediate family. Little will be said here about the larger issue of reconciling these personal goals with those of the nation or the society. Nevertheless, it is important to recognize at the outset that the success of small farm development often depends on just such a coincidence of individual and societal goals. Either the society becomes responsive to the personal goals of its individual members, or the individual is motivated to recognize the society's goals as his own.

It is fast becoming a truism among economic development specialists that farmers are "utility maximizers"; that they are purposive in making decisions that are in their best interests as they see them. It is often difficult, however, to define and quantify the farmer's concept of utility. Whether utility is synonymous with profit in the traditional sense depends on the degree to which the particular farmer recognizes and participates in commercial values. If we are to be of any relevance or use to the great numbers of farmers outside the commercial sector, we must not depend too much on the profit motive.

The pursuit of noncommercial goals may lead the farmer to make certain choices for their prestige value rather than for simple profit. He may decide to produce honey, butter, nuts, or beer in lieu of more mundane staples. He may make clothing of flax, silk, mohair, or cashmere. These items, or any surplus production, can be traded for goods and services with the farmer's neighbors—the potter, the blacksmith, the tanner, or the herbalist. The farmer's noncommercial interests may also include such things as religious gifts, the creation of artifacts, home building and decoration, the accumulation of food and goods for dowries, and the observance of births, deaths, marriages, circumcisions, confirmations, and other rituals.

Perhaps the most difficult value for an observer to appreciate in the small, noncommercial farmer is stability. The farmer values insurance against famine or crop failure; by the same token, he places a negative value on unnecessary risks. What is sometimes described as the farmer's characteristic tendency to avoid extra exertion or commitments of resources is actually a reflex attempt to assure stability. The farmer's sense of security is enhanced by his choice of the known over the unknown and by his conservation of energy and resources. The tendency of the small farmer, barely making a living under the best circumstances, to put an exceedingly high value on security and stability is a fact that must be appreciated by those who plan changes to improve the small farmer's condition.

The uses of labor

A small farmer follows a series of steps to transform his time and labor into agricultural products. These are consumed directly, sold for money, or traded for goods and services. A farmer's willingness to invest his labor in this way—to forego leisure or some alternative activity—is governed by his closeness to the minimum standard of living, as well as by cultural norms and social pressures. A farmer who is barely surviving will presumably be willing to invest his labor for a relatively low return. As his ability to produce a surplus increases, however, the farmer may well hold out for a relatively higher return before he is willing to make such an investment. At lower levels of subsistence, the only penalty for the failure of a new investment of labor is more hunger; the farmer has nothing to lose. At the surplus level, he risks the loss of his neighbors' respect.

The essential factor in small farm development is the improvement of the farmer's labor efficiency, but our ability to measure efficiency—much less to improve it—depends on our ability to comprehend the farmer's goals and his preferences for different values, his sense of the "utilities" he seeks to "maximize."

Long-term versus short-term goals

Another important aspect of the farmer's behavior in pursuit of his goals is his choice between long-term and short-term goals. Farmers at extremely low subsistence levels understandably tend to think more in terms of immediate returns and less in terms of far-future consequences. The headlong exploitation of soil fertility characteristic of traditional shifting cultivation, for example, meets the short-term survival need for food at the expense of the long-term maintenance of the resource. In contrast, the interplanting of perennial crops with annuals after the forest has been cleared and burned is evidence of the farmer's recognition of long-term considerations. Similarly, the transformation of a tropical jungle into a rubber, oil palm, or coconut plantation requires an even greater measure of long-term vision and commitment—one that the small farmer can rarely afford. Life at or near the subsistence level discourages the farmer's willingness to sacrifice a portion of his current production potential for the chance of higher production in the future. Often, however, the small farmer can satisfy both goals by gradually phasing in long-duration crops as part of his traditional crop mixture.

A family planting maize in Latin America. The oldest child leads the draft animals; the father manages the plow; an uncle spreads fertilizer in the furrow; the mother drops seed; and the youngest child kicks soil over the seed to cover it. Family labor is characterized by different levels of strength, skill, and management ability. Efficient farm enterprises use all types.

Cultural aspirations

In addition to his fundamental needs for food, clothing, and shelter, the small farmer has desires and needs determined by his culture. Ornamentation, entertainment, devotions to a deity or deities may be included in the aspirations of even the poorest farm family. The satisfaction of such aspirations will be produced at home or purchased, depending on cost, availability, and other factors. In any event, they represent goals to which labor will be diverted.

Similarly, some cultures place a great value on contemplation, meditation, and other "idle" pursuits, either by the individual or by a segment of society—priests, gurus, teachers—devoted to that activity. Where schools are accessible, the education of children often becomes a powerful cultural aspiration. The satisfaction of such cultural needs and aspirations can become an important part of the small farmer's life, despite the urgent need to attend to his vital survival needs. His farm production pattern is often designed to secure these cultural values as well as food, clothing, and shelter. In evaluating the flower offerings the farmer leaves in a temple, or the time he spends watching a historical pageant, or the value woven into a prized prayer rug, the outside observer must transcend the traditional concepts of commerce and economics, where values are measured in terms of price and quantity.

To meet his cultural goals, the farmer in an area of limited commercial opportunities may rely on his own capacity for domestic production. Even the size of his family may be expanded, at least quasi-purposefully, so that at least one daughter can be spared from agricultural pursuits to learn the temple dances; a son can enter the monastery or go to school; or a relative who knows how to weave prayer rugs can be brought into the household.

For the farmer to satisfy all his cultural goals in the commercial marketplace is clearly impossible and probably culturally undesirable. Most cultural needs will perforce be met at the expense of a certain amount of potential farm production; any successful agricultural development program must come to terms with this noncommercial fact of life. The farming-systems analysis outlined in this book attempts to give full weight to the importance of the farmer's cultural goals. The development alternatives derived from this analysis take into consideration the farmer's deeply felt need for more than the basic necessities: for participation in his community; for growing trees and plants that provide comfort, diversity, beauty, and privacy; for religious and ritual observances. Increasing the farmer's marketable food surplus so that some of these needs can be satisfied through commercial channels is only one part of a realistic and relevant development plan. Where the farmer's production potential is limited, a large marketable surplus may not be possible to achieve, and the farmer must remain dependent on his own farming system for the satisfaction of almost all his needs: food, shelter, fuel, comfort, cultural gratification, and his meager cash income as well.

Appraising development goals

The agricultural development specialist must remain constantly aware of—and on guard against—the natural tendency to superimpose his own values on those of the farmer. The reality that faces the farmer who ekes his existence from a mere half-hectare of poor land can only be understood and appreciated—and improved—if it is seen as he sees it.

This perspective may lead us to the conclusion that improved subsistence for the small farm family may be the most that can reasonably be expected, or even hoped for, in the near future, despite the creation of markets for high-value crops and other development innovations. Subsistence may be the best the small farmer can do until off-farm development offers an alternative to his half-hectare for the support of his family.

The larger farmer can more readily take advantage of market opportunities, divert his resources to the most labor-efficient, profitable crop, and spend his cash income on inputs and family needs.

4

Measuring the Well-Being
of Small Farmers

Because so many small farmers operate entirely outside the commercial sector, or very nearly so, the standards commonly applied to evaluate farm management—income, return on investment, cash flow, and the like—are inappropriate or misleading. Even the seemingly straightforward measurement of crop yield is difficult when the small farmer grows a complex mix of crops, and it doesn't mean much to measure yield when the small farmer places a higher value on reducing risk to a minimum than on increasing production to the maximum. Likewise, the indicators commonly used to measure the effectiveness of extension programs—the numbers of participating farmers or farmer contacts per extension worker—are not reliable in remote areas. Nor can they be translated readily or reliably into assessments of the overall well-being of small farm families, whose improved condition of life is the ultimate aim of agricultural development.

Need for indicators

If the success of small farm development is to be assessed in terms of the well-being of farm families, we must have accurate and objective standards that will measure the various factors that contribute to the farmer's condition. We must know if the new technologies we are developing are sufficiently relevant to the farmer's situation, and sufficiently superior to his own

Farmers who have volunteered to participate in research listen as a research worker describes a new cropping pattern.

methods, to be adopted. We also must measure our successes against the goals we have explicitly set with the farmer. In the process of devising standards and measures, we must guard against building in any implicit bias against the fact that, for the foreseeable future, most tropical farms will be small units with severe resource limitations and that subsistence farming, supplemented by the domestic production of other non-commercial goods and services to satisfy family needs, will be the major occupation of most small farmers in the tropics.

Cash flow as an indicator

In the context of noncommercial agriculture, traditional measurements of program success and farmer well-being are not only difficult to make, they can easily be misleading as well. A shifting cultivator of southern Nigeria, a Nepalese hill farmer, and a member of a Chinese production brigade may have similar cash incomes but vastly different living standards and general levels of well-being. Meanwhile, the Malaysian rubber smallholder may have a higher cash income than any of them, but he has to spend most of it to buy high-priced food.

A comprehensive index of well-being

We require standards for assessment that are objective, uniform, and clearly defined, but at the same time responsive to local values and circumstances. The following list of factors indicative of farmer well-being, originally developed for an anthropological study in the Andean region of South America (unpublished thesis by Peter H. Gore), can be adapted to noncommercial small farmers anywhere.

1. Household practices
2. Health practices
3. Artifacts and adornment
4. Home construction and furnishing
5. Agricultural wealth: animal ownership
6. Agricultual wealth: food storage capability
7. Agricultural wealth: production potential

8. Communications experience: contacts with people, transfers of information
9. Social participation: meetings and rituals
10. Attitudes

Any such list of categories, of course, must be amended and refined to fit each particular situation. In some areas, for example, adequacy and variety of diet may be a more sensitive and significant index of a family's well-being than its clothing or personal adornment.

To translate the list into a measuring tool, a 10-point scale of values must be constructed for each category. In the absence of complete and credible data from surveys, rarely available in small farm areas in developing countries, the scale of values should be calibrated by investigators who are intimately familiar with national and local norms. In addition, the categories themselves must be weighted to reflect their relative importance in the farmer's local context. Both the calibration and the weighting of the 10-point scales are susceptible to distortion by subjective judgments; great care must be taken to avoid the subtle influence of the investigator's values.

The fact that most small farmers in the tropics subsist mainly or entirely outside the commercial sector, however, should not obscure the importance of commercial agriculture as a development goal. Moreover, national and international agricultural development programs regard increased commercial activity as one index of progress, and they require data on farmers' involvement in the commercial sector for the assessment of programs which they support. In order to yield this kind of information, the list of categories for measuring farmer well-being should be supplemented with such traditional indicators as total production, yields of important commodities, cash flow, off-farm income, net income, and the fraction of total production that the farmer sells, trades, or gives away.

In conclusion, it should be emphasized that new technologies designed for recommendation to farmers with limited resources and little leeway for error must be honestly and thoroughly evaluated at every stage of development and testing. Therefore, it is of the utmost importance that we apply standards of evaluation that are appropriate to the working and living conditions of small farmers, and that can be adapted and applied by professionals working at this level of agricultural development all over the world.

5
Research in
Small Farm Development

A new kind of research—new in both direction and scope—is needed to improve the productive ability and well-being of today's small farmer in the tropics. This new research should be based on an understanding of actual farming systems and how they fit into the social and physical environments. Such an understanding is essential if research is to contribute to changing those systems in ways that will promote the achievement of the farmers' goals.

Experiment station focus

Heretofore, national research programs in the tropics have most often been patterned after those in the developed countries, and young scientists from the developing countries have been schooled in the Australian, European, or North American system. Not surprisingly, their professional work has tended toward basic research conducted at well-equipped experiment stations, to the great neglect of practical investigations under actual conditions on farmers' fields. The quality as well as the quantity of their work has been measured by the number of journal articles they have published or the number of presentations they have made at prestigious national, regional, and international meetings, rather than by the contribution their work has made to the well-being of small farmers.

As a result, it is increasingly common in developing nations to see modern experiment stations using the very latest, highly productive techniques, while literally across the fence traditional farmers continue to grow crops as they have for centuries. The striking contrast is frequently attributed to deficiencies in agricultural extension, but this view is only partly valid; the root of the problem is the lack of appropriate research.

In the developed countries, the research system is well suited to communities of reasonably well educated farmers who are served by a highly elaborated, progressive, and aggressive private agricultural industry that invests heavily in production research. A complex communication system affords a variety of channels through which farmers get abundant, up-to-date information. In addition, farmers in the developed countries often enjoy the luxury of being able to choose among a variety of alternative crops, input combinations, and investment opportunities. In such a situation, public support for a strong emphasis on basic research is justified and even essential.

Farmer focus

Small farmers in the developing countries, however, need research that is aimed directly at the practical problems of agricultural development, and attuned to the actual circumstances of their lives. The method proposed in this chapter includes a certain amount of basic research in varietal improvement, disease and pest management, plant physiology, and soil fertility. But the major emphasis is on production research, planned and carried out by and with the farmers on their own fields. This fresh approach is not a substitute for either basic research or continued technological development. Rather, it is a way of making sure that the fruits of knowledge and technology are shared with the smaller farmers, who are often excluded from agricultural improvement programs.

The outline of the research approach offered here lacks detailed prescriptions for its implementation. It is presented in this form so that it can be more easily adapted for use under local conditions in a wide variety of areas. Local adaptation is the key to the success of this approach; more exact instructions might be misleading. At the end of the book there is a bibliography of current work on this system, especially in Asia.

A list of sources of assistance for those interested in incorporating this approach into their agricultural programs appears in Appendix A.

A philosophy of collaborative research

The approach to research suggested here has important points of similarity to the Japanese system of locating testing stations in each minor political unit—prefecture, county, parish, and subdistrict. The siting of research stations is a significant factor in agricultural development, facilitating the local adaptation of research results transferred from other areas. The Japanese system has proved extremely effective in adapting national and international research findings to local farming conditions.

Two other innovative agricultural development systems have contributed to the system described in this book. One, started at the International Rice Research Institute in the 1960s, is the use of farmers' fields to test packages of seeds and materials. The other is the People's Republic of China's requirement that research scientists live and work for long periods of time with peasant farmers to gain a firsthand familiarity with their circumstances and their common sense wisdom. The former approach has had considerable success in partially commercialized areas, where increased pesticide and fertilizer sales have led to remarkable production gains. The success of Chinese agriculture, meanwhile, has increased the awareness in the international research community that the benefits of close contact between scientists and farmers accrue to both parties.

Increasing interest in these farm-centered research systems has led scientists in the cropping systems program at the International Rice Research Institute to make a thorough appraisal of the knowledge and resources to be found in traditional farming systems. In 1972, they began to document some of this centuries-old knowledge and to measure the efficiency of resource use in these traditional systems. The result of these investigations has been a greatly increased respect for the traditional farmer and a new effort to combine

Scientists know little about important mixtures like rice and cabbage. Such combinations allow farmers to use their land and labor more efficiently and may lower risk.

traditional farming knowledge and skills with the trained insights and experimental method of the scientific researcher.

The first experiments in farmer-scientist collaboration began in 1973. Trials were planned with the farmers and planted on 0.1-hectare plots on the farmers' land, using materials provided by the research station. Farmers and researchers regularly visited the fields together, but the day-to-day management and care of the plots were the farmers' responsibility.

Five years of these trials demonstrated their value. The results rapidly expanded scientists' understanding of the systems, and scientists and farmers developed a mutual respect that has benefited both. The farmers showed a readiness to adopt the most successful experimental technologies from the trials. Certain requisites for the success of the new research system emerged from this experience:

• A thorough site description, including good environmental data on soils and climate, is the first step toward the selection of appropriate technologies to be tried. This data can be used later to extend the experimental results to other areas with similar conditions.

• The research staff should have farming experience and competence. Farmers will not respect anything less.

• An attitude of cooperative learning must be maintained by both farmers and researchers. The pedantic, teacher-student attitude characteristic of many extension programs has no place in this collaborative research effort.

• The farmer must be part of the research team, involved in making plans and decisions at all levels and stages and sharing credit for results.

• The farmer should not be paid in cash for his participation. Materials may be provided by the program, however, and he may be guaranteed some compensation in kind if the experimental plots fail completely.

• Participating farmers must be carefully selected. Once chosen, they should be left free to make their own production decisions and to do the work themselves with help from their families. Middle-aged or older farmers are preferable because of their long experience, and because they are less likely to be kept on the edge of subsistence by the demands of a growing family. Such farmers are likely to be more inclined to experiment, and the research plan should encourage this tendency in every possible way.

• Constant contact should be maintained between the farmer and the researchers. Daily visits to the field should be made by a junior researcher in the company of the farmer; weekly visits should be made by the senior researcher.

• It is helpful to base the research team at an experiment station where scientists are engaged in both basic and practical developmental research. The resulting cross-fertilization of ideas and insights will be valuable for both scientists and researchers.

• Extension workers can be brought into the research team after it has been in successful operation for at least a year.

• In order to demonstrate the value of the new approach, it is important to include in the program enough farmers to allow meaningful statistical comparisons of experimental results.

The involvement of scientists with farmers that is the hallmark of this approach capitalizes on their mutual motivation for improvement. Enjoyment and pride in experimentation, shared by both groups, is an important force for getting the program under way, and then for carrying it through to a significant conclusion. In the process, the farmers will learn the science of precise measurement and comparison, while the scientists will gain insight and experience in the real world of the cultivator.

Nevertheless, it is not necessary for all scientists to participate directly in on-farm research. The excesses of the Chinese system, which sends every scientist to the farmers' fields for a full year, should be avoided by a process of rational selection. Production agronomists, specialists in pest control, soils, and farm management, and social scientists such as sociologists and anthropologists will benefit greatly from on-farm experience. Plant breeders and physiologists can more profitably devote their time and efforts to the basic research that must be conducted under the carefully controlled conditions of the experiment station and the laboratory.

On-farm research

The methodology we suggest for on-farm research develops in a logical sequence of steps:

1. *Selection of the target area.* The methodology will be most effective in an area where small farmers make up an important segment of the population. Observation and detailed description of existing local farming systems will require socio-economic surveys and technology inventories if these do not already exist.

2. *Description of the environment.* Collection of complete and accurate data on climate, soils, and other salient aspects of the physical environment is crucial. Depending on the availability of good existing data—a relatively rare circumstance in most developing countries—this process may involve only a quick inventory or the painstaking collection and analysis of original weather data, surveys of local farmers, aerial photography, soil classification, and mapping.

3. *Design of alternative technologies.* Working closely with the selected farmers, the scientists plan what tests can be done to accomplish specific mutual goals with the available resources. The individual goals of each farmer can be described on an objective scale such as the one suggested in Chapter 4, and a precise list can be made of his individual resources, including their special characteristics and constraints. The range of possible alternative technologies is determined by the scientists, based on their knowledge of the area and its production potential. The farmer, however, should have the last word on what innovations will be made on his land. Both the farmer and the scientists should estimate the effects they expect each technology to have on crop production.

4. *Testing the new technologies.* The joint planning of experiments by scientists and farmers should include agreement on timing and supervision, as well as on provision of seeds, plants, animals, and outside inputs, and on the care of the plots. Daily supervision and care is vital to the test, so that problems can be solved promptly and good records can be kept. If a large number of trials are involved, a junior researcher may be given the responsibility for the daily visits, with weekly visits

by the senior scientist. On all visits, however, the farmer should accompany the observer.

5. *Evaluation and refinement.* As the harvest is completed, or as animal production reaches a sustained level, the scientist and the farmer should collaborate on a joint evaluation of the trials. At this stage, it is especially important that full weight be given to the personal values and goals of the farmer, for whom even a spectacular yield increase can entail problems that the scientist cannot discern. In the Philippines, for example, a farmer who multiplied his cash income 15 times by planting disease-resistant tomatoes was subjected to such social pressures from his less successful family and neighbors that he declined to plant them the following year. In this case, spectacular commercial success was personally unacceptable to the farmer; a sustained increase in food production for consumption at home would have suited him better. Adjustments in the experimental design can be made at this stage to satisfy the reconciled evaluations of the first-trial results by the farmer and the scientist.

6. *Continuation of trials.* Most experiments must be repeated several times over several seasons to demonstrate the adaptation of new technologies to the varying agroclimatic conditions of the target area. The scientist should let the farmer decide whether to increase the size or the extent of the experiment. The farmer's judgment becomes part of the scientist's evaluation of the experiment's value. If the experiment yields positive results, extension workers may be enlisted for the second and third trials so they can learn the new system and its results and integrate it into their future work with farmers. It is important to avoid bringing in extension workers who are so wedded to established technologies and approaches that they will resist any innovation, no matter how demonstrably valuable.

7. *Final evaluation.* A final evaluation of a new technology can only be made on the basis of several seasons' experience. (The scientist's need for long-term evaluation, however, should not, and probably cannot, prevent early adoption of an innovation that farmers perceive to be beneficial.) The changes brought about by innovation must be measured against the

original goals of farmer well-being established before the experiment began. Gains in some areas may be offset by losses in others. If food-grain production is expanded at the expense of legumes, for example, the protein available to the family may actually decline in both quantity and quality. Such a paradoxical effect would not show up if well-being were measured solely in terms of farm income or total food production. Additions to the farmer's household are readily measured, but the evaluation must also take into account more subtle changes in health, community contacts, communication with outside influences, and aspirations for children. Such changes can only be evaluated if careful and sensitive baseline measurements have been made on a well-defined scale before the beginning of the experiments.

When the final evaluation has been completed, the successful changes in the farmers' traditional systems can be extended to other farmers in the area through normal extension channels. Moreover, the same technologies and systems can be confidently introduced for local adaptation and trials in other areas with similar agroclimatic and cultural conditions. This approach is based on critical elements that are largely or entirely omitted in most current systems of agricultural research and development:

- Detailed classification of environmental factors
- Collaborative planning and management of trials by scientists and farmers
- Evaluation of experimental results with the farmer and in terms of the farmer's goals

The farmer's actual participation in the planning, execution, and evaluation of research should be clearly distinguished from mere research in farmers' fields initiated and controlled completely by scientists. The latter approach simply provides a test of technological components in various actual farm environments. The results may be valuable to the scientists, but they do not show how well the new technology performs under the farmer's management, nor how it integrates into his farming system. And they do not encourage the adoption of successful innovations by the farmer-participant. It is crucial that the research organization appreciate the value of joint farmer-scientist planning, testing, and evaluation of technological changes. The farmer's criticism or rejection of the researcher's favorite methods or new varieties is often difficult for the researcher to accept. It involves both his personal and his professional pride. But if the farmer's opinion is ignored, discounted, or even ridiculed, the fragile connection between farmer and researcher on which this entire system depends will be broken.

Part 2

Critical Factors
in Small Farm Development

6
Physical Limits
to Cropping Intensity

The potential cropping intensity of any farming system depends on several physical factors that the development planner must take into account in estimating the potential production of the environment. We will outline the limiting factors here, giving detailed examples of the most important ones.

In any country, district, or even village, the limiting factors—the availability of water, for example—are variable and complex. The planner must bring order to this variability and complexity in such a way that he can use the environmental data to guide agricultural development efforts, to interpret research results, and to extrapolate those results to similar areas. The planner's system for classifying environmental data, to be truly useful, must have certain characteristics.

First, the classification must stand by itself, not be specific only to a particular location.

Second, all limiting factors in the environment must be treated as having continuous gradations from one location to another. Most factors, such as rainfall, temperature, and physiography, are continuous by nature. Others, however, such as soil type, vary sharply from place to place and are more difficult to classify.

Third, in measuring each factor, care must be taken to include only those aspects that have a significant effect on crop growth and management.

Fourth, the description of weather, a prime factor, must be

quantitative, permitting the planner to calculate statistically the probabilities of occurrence and ultimately to determine the farmer's risk.

Fifth, the description of factors in the physical environment must be simple enough to permit ready identification and rational selection of areas for development.

Sixth, those limiting factors that change gradually over relatively large areas should be mapped. Others—such as flood depth or tillage capability, which change from field to field or even within a single field—need not be mapped. Although detailed mapping of these latter changes is probably too difficult to justify the effort involved, for planning purposes it is important to know the magnitude of the changes in each location because production capability is seriously affected.

Recognizing the variability and complexity of the natural environment, the planner evaluates the total target area for its agricultural production potential. Except in the case of relatively large areas that have gradual environmental gradients, however, planning for each field must be done by the local extension agent working closely with the individual farmer, who is the best possible expert on such highly localized environmental factors as depth of flooding and tillage requirements. Together, they can readily determine the classification of a particular field for each salient environmental factor, and thus arrive at a reasonable estimate of the field's crop potential.

An example: Water

We will illustrate the operation of the classification system by looking in some detail at one limiting factor: water availability. The major sources of water are rainfall, irrigation, soil moisture, and flooding. While the pathologist may also find it essential to know about humidity, and other specialists may need measurements of still other water sources, these four are the primary elements of total water availability as it affects cropping potential. Whatever its source, the availability of water is measured as the amount of moisture per unit of time. It makes little difference to the crop whether the water comes from irrigation or from rainfall, except that if the source is highly variable, considerably more water will be needed to ensure a sufficiency.

As an example of how a classification system works, we will look at a scheme proposed by the International Rice Research Institute (IRRI). To classify water availability in rice-growing areas of the humid tropics, IRRI decided to make a basic distinction between those with more than 200 mm of rain per month and those with less. The 200 mm level corresponds generally to the water requirements of rice grown in submerged soil in paddies. Whether this level corresponds exactly to water requirements in a particular area, however, is not a critical question; 200 mm is a useful benchmark that can be correlated with crop growth and management potential. The 100 mm level, and even the 50 mm level in drier areas, may also be important subcategories of the 200 mm classification.

In most cases, monthly rainfall data are adequate for classification purposes. They also have the advantages of being available in summary form for long periods in most countries and of being less variable than data from shorter intervals. For classifying other factors, however, weekly data are more useful because they correspond conveniently with the common time interval for farm management studies.

Figure 3 illustrates the main water-availability categories used in this classification system. Subject to special conditions, each category corresponds to a specific rice-growing potential:

• *Category I:* Transplanted rice can be grown in these areas, which have less than three months with 200 mm of rain, only if the soil puddles easily (puddling is cultivation of wet soil to break down soil structure) and thus is highly impermeable to water percolation, and if runoff water from higher fields or paddies is available. Production is risky.

• *Category II:* These areas, with 200 mm of rain for three to five months, are the prime areas for growing a single crop per year of transplanted rice.

• *Category III:* In areas with 200 mm of rain during five to seven months, two crops of early-maturing rice can be grown in paddies. Unless the rainy season begins abruptly, the first crop should be direct-seeded on soil that has not been puddled.

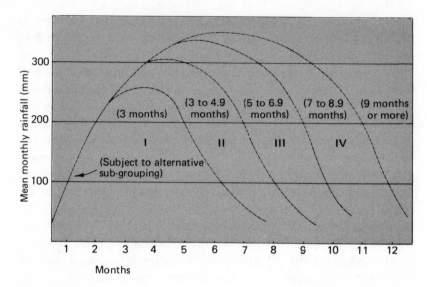

Figure 3.
Categories in a rainfall classification for rice-growing areas of Asia.
(Number of months with over 200 mm of rainfall are shown in parentheses.)

- *Category IV:* With from seven to nine months of 200 mm rainfall, two crops of transplanted rice can be grown.
- *Category V:* Areas with more than nine months with 200 mm of rain can support continuous rice production.

In addition to water availability, several other crop management factors limit rice production:

- *Tillage:* The tillage characteristics of the soils commonly used for upland rice require 100 mm of rain per month for good seedbed preparation, unless large, tractor-mounted, rotary tillage equipment is available. This water requirement applies to upland field crops as well as to direct-seeded rice on dry soil.

For converting from upland to paddy rice, 300 mm of rain per month is required; for converting from paddy to upland rice, on the other hand, the rainfall must be less than 100 mm per month, with the exception of sandy, river-levee soils, which can be converted under conditions of higher rainfall. If the rainfall is more than 200 mm per month, however, mechanical cultivation for weed control in unpuddled soil will be ineffective in many seasons.

- *Planting requirements:* Direct-seeded rice on unpuddled

soil requires at least 100 mm but not more than 200 mm of rain per month at the time of planting. Higher rainfall saturates the soil for prolonged periods, reducing seed germination.

• *Harvest requirements:* On small farms, rice can be harvested with up to 300 mm of rain per month if unflooded land is available where the rice can be spread to dry. In areas with more than 300 mm of rain per month, however, the rains are too frequent to permit sun drying, and mechanical drying is required.

To make the fullest use of the growing season, other crops must be grown in rotation with rice. The theoretical cropping potential of the land can be approached with the addition, in rotation, of other crops with differing rainfall requirements. Among the most likely are maize, sorghum, cowpea, mung bean, sweet potato, groundnut, and soybean. Among them, these seven represent widely differing environmental requirements. They also are, after rice, the most widely cultivated food crops in the humid tropics. Theoretical and empirical determinations of the yield potential of these crops can be used to estimate the performance of many other crops with similar environmental requirements. (For Burma and Bangladesh, jute and sesame should be added to the list of alternative field crops; for Latin America, black bean should be added.)

The water requirements for these crops are generally as follows: All require 100 mm of rain or more per month. If the crop, such as maize or sorghum is planted in a soil containing less than 10 percent available water, 150 to 200 mm of rain per month are required for the early stages of growth. At planting and for the first two weeks of growth, maize, mung bean, sweet potato, and cowpea will tolerate the broadest range of moisture, generally doing well with anywhere from 10 to 80 mm of rain per week. Sorghum should have between 10 and 50 mm of rain per week, and soybeans and groundnuts not more than 40 mm.

Such abundant rainfall makes good soil drainage a critical factor in the success of these crops. In general, soil drainage and moisture-holding capacity are more important at planting time than during the harvest. In the Philippines, maize can be harvested as dry grain during the monsoon period, with up to 300 mm of rain per month (up to 50 mm per week at a 0.5

probability). Sweet potato, mung bean, soybean, groundnut, and especially sorghum must have less than 100 mm of rain per month (10 mm per week at a probability of 0.5) in order to be harvested with current small farm methods.

Mechanical grain dryers would increase the rainfall range at harvest considerably for maize and rice, but not for sorghum and the legumes, which will not tolerate the high moisture levels in the weeks preceding harvest even if mechanical dryers are available for the harvested grain. Many sorghum developments in the Philippines, Thailand, and Vietnam have failed because this fact was ignored.

Cropping pattern potential and water availability

The province of Batangas in the Philippines can be used to illustrate the use of the rainfall classification system to choose among the possible alternative crops to rotate with rice (Fig. 4). The rainfall in the area falls in Category III: 200 mm of rain per month for five to seven months. Indeed, approximately 25 percent of the Philippines and 75 percent of Bangladesh fall into this same rainfall category. The mean temperature varies from 18°C to 23°C and is not a limiting factor in cropping potential. The soil is a well-drained clay loam, representing no physiographic limitation for upland crops, at least in eastern Batangas.

Thus, the determination of the area's cropping potential, for rice as well as for the rotation crop, is a relatively straightforward matter. Since the soil percolation rate is too high and the water table is too low to allow puddling without irrigation, rice can be grown as an upland crop, direct-seeded on unpuddled soil. Because puddling is not possible, what might be a two-crop rice area under lowland conditions becomes a double-cropped upland area.

Figure 4 shows the appropriate planting dates for those crops whose yields, under good management, can be expected to attain 50 percent or more of the best yields under experiment station conditions. Where the potential for commercial yields falls below this standard, the crop is deemed unsuitable for the area. These planting dates for Batangas, derived from the

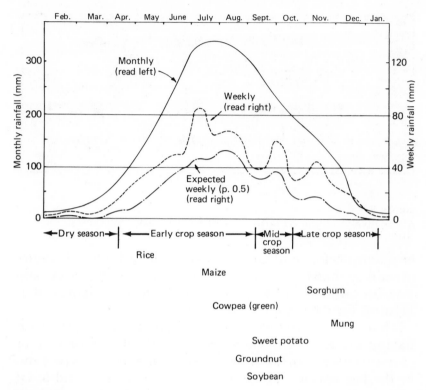

Figure 4.
Water availability and field crop patterns for upland rice farms of eastern Batangas, Philippines.

rainfall classification of the area, have been validated by three years of on-farm trials under farmers' management.

Of the early, middle, and late growing seasons, the midseason allows the fewest alternative crops, while the late season, with two months of 100 to 200 mm of rainfall, allows the most. Rice cannot be planted after June because the heavy rainfall makes direct-seeding difficult and because rice that matures later than September produces low yields. Maize, on the other hand, can be planted and harvested in any season, but green harvest in September is the easiest. Maize planted after November is used mainly for fodder, and the dry stalks are stored to feed animals in the dry season, when grain production is limited. The maize intended for dry stover and storage cannot

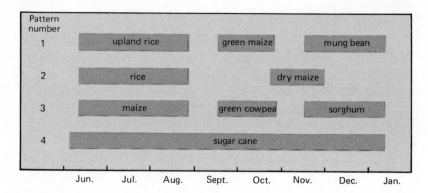

Figure 5.
Alternative cropping patterns for eastern Batangas, Philippines.

be planted before mid-October because of the high probability of rain at harvest. Sorghum, groundnut, mung bean, and soybean cannot be planted earlier than the dates indicated in Figure 5 for the same reason.

Thus, the alternative cropping patterns appropriate for Batangas are: in the early season, upland rice, maize, or cowpea; in the midseason, sweet potato, maize, or cowpea; and in the late season, the crops shown in Figures 4 and 5. Of course, other crops are possible. Sugarcane, for example, is an alternative. Jute or kenaf would grow well, but probably would not compete economically. Other vegetables that would grow well in the Batangas environment would be vulnerable to the heavy rains of August and September.

The Batangas model will work equally well in other areas of the Philippines where temperatures and soil drainage conditions are not limiting factors. Place-to-place differences in the amount and duration of rainfall simply alter the lengths of the early, middle, and late planting periods. Thus, the crop potential for most of the upland rice areas of the Philippines can be estimated quite accurately despite local variations in rainfall, and the multiple-cropping options for these areas are known.

From among these options, the particular cropping pattern chosen by the individual farmer depends on several economic and management factors, which are discussed in detail in the following chapters. Before these factors come into play, however, the land must be classified according to rainfall. As the Batangas case illustrates, the characteristics of the rainfall classification are:

- It is discrete. A particular area may fall in one category on the average, but in any given year it may be classified quite differently, and it is this immediate classification that is the operative one for the farmer. Insofar as it depends on rainfall, the crop potential of any area would be the same as for other areas having the same rainfall in that year, despite other differences among the areas. Crop production for a particular area in a given year may depend more on the area's rainfall classification than on its location.

- Rainfall classification is continuous. In mapping, there must always be a Category III area between Category II and IV areas.

- Quantitative measurements in the classification system make it possible to calculate the frequency and probability of rainfall.

- The limited number of categories in the classification system can be correlated with major differences in cropping potential.

For lowland areas where the soil can be puddled, similar water-availability classifications can be made by adding water accumulation as an additional factor.

For lowland rice, transplanting usually requires three to four weeks of standing water prior to transplanting. Water should remain in the paddy until two or three weeks before harvest. Direct-seeded rice should have standing water in the field within two to three weeks after planting if the soil is puddled, and within four weeks if direct-seeding was done on unpuddled soil. For planning purposes, these planting requirements should probably be considered as minimums for rice cultivation. As a rule of thumb, flooding should have an 80 percent probability of occurring at the desired time, leaving the field without standing water for no more than one week at a time, before planting should begin.

Transplanted rice requires at least two weeks with water standing in the field before transplanting to allow for thorough land preparation with current Asian methods. If there are too many weeds or other organic materials in the field, this preplanting period of flooding may extend for as long as a month. A similar flooding period is required before trans-planting if the soil is low in pH and high in iron compounds: the chemical dissipation of excess organic acids and harmful iron compounds in submerged soils will prevent the inhibition of plant growth. The standing water should remain in the paddy until two weeks before harvest to prevent loss in yield.

For direct-seeded rice, present methods are barely adequate to control weeds for the first 30 days without flooding. It should be assumed, then, that no longer than 30 days should elapse from the time of planting until the first flooding of the field.

The hypothetical rainfall and standing water pattern shown in Figure 6 is similar to that of a portion of the Central Luzon rice-growing area in the Philippines, which has more than 200 mm of rainfall during four months of the year—enough that there is standing water in the paddy from two and a half to four months of the year. The duration of flooding depends on the farmer's decision about when to drain his field (assuming that the farmer has a high-side paddy, as described below).

In a year of average rainfall, this area would have the potential for a single crop of rice and a limited potential for alternative upland crops. Recent attempts in this area to grow two crops of rice without supplemental irrigation have been less than successful because of the limited water supply. Wetter areas—those with at least five months of more than 200 mm of rainfall and five months with standing water in the paddy— have the potential for two crops of early-maturing rice. These conditions are common in rainfed rice-growing areas.

Crops preceding rice in the rotation must mature in 70 days or less and must be harvested in months with 200 mm or more of rain. These conditions severely limit crop potential for the average farmer. Most farmers are unable to control the drainage of their fields, which is essential for growing crops before the rice crop.

Following the rice crop, the most likely crops are the drought-tolerant legumes, which require a minimum of land preparation and applied nitrogen. Until rainfall has dropped to less than 100 mm per month, no upland crop can be planted

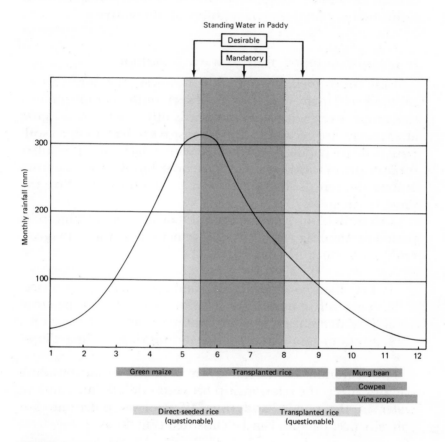

Figure 6.
Possible crop patterns in lowland rice areas having three to five months
of good rainfall.

after rice on the typically heavy rice soils of Asia, except possibly on sandy, river-levee soils. Vine crops, such as watermelon, cucumber, and squash, can be grown after rice if at least a minimum of supplemental irrigation is available. On lighter soils, cantaloupe is a potential alternative.

Importance of topography to water conditions

With supplemental irrigation, however, flooding can be prolonged in many cases for five or six months, permitting two rice crops. Even with irrigation, it is difficult to grow maize after paddy rice because maize is sensitive to waterlogged soil. Irrigated rice paddies—even those with ridges—usually suffer some degree of waterlogging. As a crop to follow rice, sorghum, among the cereals, has a greater potential than maize on the more fertile fields.

Each individual paddy should be classified according to its period of standing water and the farmer's ability to manage it. Only a few categories are required:

1. Less than two months of standing water   (no rice)
2. Two to five months of standing water   (one crop)
3. Five to eight months of standing water   (two crops)
4. Eight or more months of standing water   (three crops)

The elevation of the paddy in relation to surrounding paddies affects the relationship between rainfall and standing water in the paddy, and thus affects crop potential. For convenience, paddies can be classified (Fig. 7) as:

1. High interior: no surface movement of water in or out is possible with normal rainfall.
2. High side: no water normally drains into the paddy, but water can be drained out.
3. Intermediate: water normally drains in and out.
4. Low: water normally flows into the paddy, but the paddy cannot be drained.

A paddy that receives runoff water from those above it

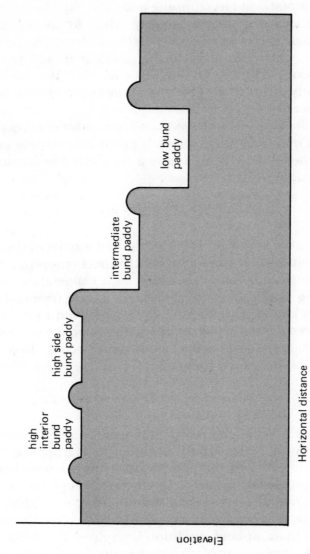

Figure 7.
Physiographic classification of rice paddies.

during the occasionally heavy early rains will retain standing water well before the higher paddies, even though it receives the same rainfall, and its cropping potential will also be different. The paddy's drainage capacity also affects its cropping potential. Heavy rains in the early and late growing seasons require surface drainage of the paddy if its soil is to avoid saturation. Thus, both high-interior and low paddies are generally unsuited for growing rainfed upland crops before or after rice.

Besides water-related conditions, a number of other physical factors may determine an area's potential cropping capacity. Nevertheless, in any area only a relatively few factors are of prime significance.

Temperature

At higher altitudes, or at the higher latitudes of the tropics, low temperatures can significantly affect cropping. In drier, lower areas, on the other hand, high temperature may be a limiting factor. Crop tolerances to extreme temperatures are usually known to farmers in these areas, and rules of thumb have been devised to calculate local crop potentials. Except in steep mountainous areas, temperature zones, like rainfall zones, can be readily mapped.

Tillage capability

In areas with clay soils and heavy rainfall, the tillage capability of the land limits the intensity of cropping. We have already mentioned that after a rice crop has been harvested, many puddled soils cannot be converted to upland crops if the average rainfall is more than 100 mm per month (about 10 mm per week at a probability of 0.5). Indeed, some soils cannot be worked in an upland condition with rainfall over 30 mm per week at a probability of 0.5. On the other hand, some soils, like Mahaas clay in the Philippines, have such rapid internal water movement that lowland rice cannot be grown without irrigation where the water table is below the soil surface. Upland rice can be grown for September harvesting, however. With five months of rain above 200 mm, two crops a year could be grown, were it not for the high rainfall in September and October that makes tillage for the second crop extremely difficult. Tree crops or sugarcane, which do not require midseason tillage, are more suitable for the climate in this area. Irrigated fields, or those with a water table high enough to permit water to be held in the paddy, are better suited to lowland rice.

Tillage capability is extremely important in a tight cropping sequence that requires quick turnaround between crops, precise cultivation, or extensive seedbed preparation. There is a pressing need, therefore, for a classification system that can be used to describe the tillage capability of all soil types. Such a system should be based on the soils' lower limits of plasticity, which are easily determined.

As an alternative basis for classification, the change in bulk density that occurs with puddling is an accurate index of tillage capability, although it is much more difficult to determine. The bulk density of montmorillonite clay such as Mahaas, for example, is up to 15 percent greater in its dried condition after puddling than in its dry upland condition. Therefore, the soil's structure is easily lost if it is worked under any but a narrow range of moisture conditions. To be converted from puddled to unpuddled use, it must be tilled as it dries. Then the hard clods can only be broken down by the shrinking and swelling that occur during one or more cycles of wetting and drying.

Other clay types, as well as coarser textured soils, undergo much less change in bulk density when they are puddled, perhaps as little as 3 to 4 percent. These soils can be tilled when they are relatively wet, and they can be converted from puddled to upland use when they have high moisture content.

For the development planner, a combination of an area's tillage capability and rainfall classifications would provide a reliable guide to its crop management potential under different cropping intensities. It is essential that such useful indices be developed as soon as possible.

Soil fertility

Soil fertility limits crop potential under certain conditions. Soils that readily support a single crop of lowland rice, for example, can be deficient in several micronutrients if upland crops are added in rotation. These shortages must be made up before upland crops can be grown successfully. In iron-deficient soils, for example, flooded rice grows acceptably only because iron is more available to the plant in submerged soil. On the older, highly weathered rice soils of mainland Southeast Asia, a single crop of rice per year can be grown under lowland conditions with the addition of modest amounts of fertilizer. Studies in central Thailand have shown, however, that upland crops following rice on these soils require much greater applications of nutrients. On many of these soils, therefore, costly inputs are required to grow cereal crops such as maize or sorghum after lowland rice, even with irrigation. Legumes, which provide their own nitrogen, can be grown more economically.

In other areas, low cation-exchange capacity makes soils difficult to manage under high cropping intensity in upland conditions, because the concomitant high fertilizer input can cause the soil's pH to shift from season to season. Multiple cropping in northern Thailand has been hampered by this problem.

Other factors—light intensity, for instance—certainly influence crop productivity. As yet, however, the operation of these factors is not sufficiently understood, and the evidence is insufficient to permit their use as predictors of crop performance or criteria for choosing among alternative crops.

Environmental classification systems

The definition of environmental categories as a guide for agricultural research and production programs can be greatly facilitated by existing reliable survey information from a number of sources. In Bangladesh, for example, the United Nations Development Program–FAO 1971 soil survey, "Agricultural Development Possibilities," is an excellent source of useful data.

Whatever their source, the categories established by the development planner should, first, be realistic in their utility for the measurement and prediction of environmental variables, and they should be precise to a practical degree. Second, the complexity of the categories should be commensurate with the sophistication of the national research effort in dealing with varied environments. A highly structured and well-developed research organization can deal effectively with a relatively fine-gauge breakdown; a less sophisticated organization probably cannot. Finally, the choice of categories should be responsive to the needs and concerns of national agricultural production programs.

The classification system should divide the target area into the smallest practical number of crop-potential zones. Detailed soil and hydrological surveys are invaluable in defining zones as well as in designing the classification categories, and in locating other areas to which research results from the target area can be extrapolated.

Thus, the identification of increased cropping potential depends not only on distinguishing different crop environments, but also on knowing the responses of various crops to those conditions. Even without detailed local data on crop responses, however, the development researcher or the farmer can operate at first on an intuitive understanding of the requirements of a few typical crops. In time, the accumulated experience of different crops' responses to different environmental conditions will add up to a more comprehensive and sophisticated understanding.

The response of mung bean to rainfall, as shown in Figure 8, is an example of observed crop response to a prime environmental factor. Mung beans were grown, using the best available varieties and cultural practices, under both supervised farmer management and research management. Yields proved to be optimum with between 50 and 100 mm of rainfall during the 70-day growing period. Higher rainfall usually produced lower yields, and seed quality declined if rain came at harvest time. Mung beans can thus be planted only as the rains taper off at the end of the wet season. (In Thailand, however, the same rainfall effect occurs at the beginning of the season because of a bimodal rainfall pattern that results in a short planting season followed by drier weather before the onset of the major monsoon.)

The identification of alternative cropping patterns that will more fully utilize available production resources requires a

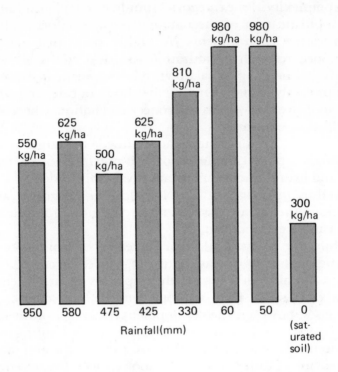

Figure 8.
Mung bean responds to low rainfall: yield in relation to rainfall between planting and first harvest (data from trials in Batangas, Philippines).

new kind of collaborative contribution from the various specialists involved in agricultural development. The agro-climatologist, the soil physicist, and the soil classification expert, for example, must carefully structure the diverse farming environments to identify crop potential and to focus on research and development. Similarly, the production agronomist must become more sensitive to the environmental requirements of alternative crops, focus his research on significant agricultural environments, and identify the environments to which his research results can confidently be extended.

7

Economic Determinants of
Crop Type and Cropping Intensity

So much has been written about farm management that another analysis is hardly needed here. It is important nonetheless to review a few key concepts bearing directly on the subject of cropping intensification, on which further economic research is most needed.

The agricultural potential of a given piece of land is largely determined by the physical factors—water availability and the rest—mentioned previously. Besides these, however, social and economic factors—such as the availability of labor, power, and cash to purchase inputs—also determine cropping potential. Still other factors—including markets, technical services, and the availability of inputs—are essential, but these are outside the scope of this book.

In order to know what kind of intensive cropping technology is most appropriate to a given farm, it is necessary to determine certain basic facts about the farmer's socioeconomic situation: How much labor is available to him? Does he have—or can he rent—animal power or mechanical power? Is his potential cash income adequate to buy commercial fertilizers and other chemicals? Does he have the necessary management skills, either himself or in his family, to run a more intensive agricultural enterprise? The availability of these resources can readily be measured on the scale of growth stages outlined in Part 1 of this book. Because different intensive-cropping technologies have very different requirements for farm resources—labor, power, etc.—an understanding of the particular farm in terms of these essentials is crucial to the selection of the most promising cropping system.

These socioeconomic factors can be divided for convenience into those that have a direct, primary influence on cropping potential and those which affect cropping potential indirectly through their influence on primary factors. The five primary determinants of cropping-pattern potential—labor, management capability, power, cash inputs, and markets—will be considered first.

Labor

The availability of labor imposes a major limitation on the crop type and the cropping intensity that a farm can support. It is obvious that different crops and crop combinations respond differently to labor. An increase in labor means greater planting precision, better weed control, and more timely and complete harvesting. The crop responds initially with increased productivity, until the yield tends to level off as the crop approaches its maximum potential production. At this point, additional labor ceases to produce increases in productivity, and the marginal return on labor declines. Differences in the return on labor may be considerable for different crops and crop combinations.

Farmstead areas that use the "three-story" system described in the following chapter require relatively little labor. When these areas are properly farmed their productivity can be high, but they do not respond to ever-increasing inputs of labor. For example, black mung bean, commonly seeded by broadcasting onto a prepared bed, is highly competitive with weeds and normally requires little additional labor until it is harvested. Green mung, on the other hand, is slightly less competitive with weeds, requires more than one harvest to achieve optimal productivity, and will respond to weeding under many conditions. Most field crops are not as competitive with weeds, and they respond to additional labor as maize does, rewarding precision planting, weeding, and pest control.

Vegetables are still less competitive. They may require special seedling care, such as labor-intensive seeding in nurseries and later transplanting to the field. They may also

A husband and wife transplanting millet. Small-scale farmers can raise productivity through methods that, on a larger scale, are impractical.

require mulching or irrigation, and several weedings and harvestings. Since they usually have a high market value, however, vegetables can often justify intensive labor. Many intensively farmed intercrops fall between field crops and vegetables in their return on labor.

Some crops will give moderate returns, but only to low levels of labor. Field crops such as maize and rice yield moderate returns, but at much higher levels of labor. The return on labor depends not only on the crop, but also on the physical environment. If other factors, such as moisture, become limiting, the potential increase in yield probably will not be high enough to justify the expenditure of much labor on weeding or harvesting.

In a labor-surplus situation, it is often considered advisable to encourage the use of labor-intensive technologies. As a national policy, such an approach may well be consonant with an overall social purpose of increasing total farm production, but from the individual farmer's viewpoint, the return on his added labor is worthwhile only if it yields farm products or income that he wants. As discussed in Chapter 3, the farmer pursues his own goals, which may or may not coincide with larger political or social purposes.

Management capability

Management capability is an often overlooked resource that is closely related to labor availability. The management of farm production includes all production-connected activities that cannot ordinarily be performed by common farm laborers. The total of human resources on the farm is thus divided into management and labor functions, although on a small, family-operated farm both functions are performed by the same people. Management involves making decisions, performing certain technical operations requiring exceptional skills, and supervising other farm operations when necessary.

High-value crops, such as vegetables, demand rather meticulous care in growing and marketing. Land use planning, the securing of seed, planting, pest and weed control, harvesting, quality control, and marketing all require close, active management. In small farm agriculture, where wage rates for outside labor are low, the available labor supply tends to be unskilled, sometimes undependable, and even untrustworthy. Field laborers cannot be held responsible for making judgments or decisions about the operations they are hired to perform; thus, the farmer and members of his family share management responsibilities on all but the largest farms.

On a family-operated farm, then, cropping intensity may be limited by the management capabilities of the family even though off-farm labor is available. The size of a family farm may be limited to a maximum of about 2 hectares by the management requirements of intensive vegetable cultivation, and cropping intensity will usually begin to decrease as farm size exceeds 1 hectare. If off-farm labor and power are available, however, intensive crop sequencing can be found on farms of up to 10 hectares. As the limits of other resources are approached, their optimum use must be ensured by adjustments in cropping intensities and patterns, and the need for careful management increases. Therefore, on small farms the intensive use of such physical resources as land and water depends on the farm family's commensurate ability to furnish management services.

Power

Power is another major resource that has a direct effect on cropping capability. In the production of cultivated crops, essential seedbed preparation involves loosening the surface soil layer to disrupt weed growth, burying the debris to some extent in the process. This process is called primary tillage. There is a pronounced inverse correlation between the energy invested in primary tillage and the energy required for secondary tillage or chemical control of weeds later in the growing season, when the crop is well advanced. For either primary tillage or weed control, the farmer can choose to use human labor, animal or mechanical power, chemicals, or some combination of these. His choice depends to a large extent on the resources at his disposal.

Primary tillage by hand yields extremely low returns on labor unless the crop is extremely valuable or it requires little tillage. A farmer who lacks mechanical weeding equipment or access to herbicides must therefore choose a cropping system requiring relatively little weed control, such as a tree crop; or grow a high-value crop, such as a vegetable; or accept the extremely low productivity of hand labor.

On the other hand, mechanical equipment can be used most effectively in a cropping pattern in which primary tillage and inter-row cultivation use a large part of the total labor required to grow the crop. Field crops grown in rows, such as maize, use a far higher ratio of mechanical power to human labor than do vegetables or intercropped field crops.

If supplies of both mechanical power and human labor are limited, chemicals can be substituted for weed control, or the farmer can grow crops that need less labor. The upland rice-maize-cassava intercrop of Indonesia is one such combination.

Primary tillage is done during the digging of the previous cassava crop in the dry season. Rice and maize are planted at the start of the rains, with cassava following one month later. This crop combination forms a rapidly closing, dense leaf canopy that shades out weeds. The maize is harvested first, followed by the rice. After the rice harvest, the cassava closes in for the final six months of growth. In a situation where mechanical power is limited, this system takes advantage of a ten-month growing season, returns three crops from a single primary tillage operation, and requires little weeding, yielding a maximum return on human labor.

Power-assisted primary tillage is necessary, however, in an intensive sequence of field crops. Animal power has the advantage of being well suited to use on heavy soils with high moisture content, where it is also more versatile in inter-row cultivation. If the draft animal is also considered as a source of milk—or, ultimately, of meat—it becomes even more cost-efficient as a power source.

On the other hand, an animal's feed requirements can complicate the cropping pattern on a small farm, and the maintenance of draft animals also requires considerable labor. While machinery requires less maintenance labor and provides more rapid field operations, it also requires greater capital investment and continuing expenditures for its operation.

For intensive cropping sequences requiring rapid turn-around between crops, a source of power and the ability to use it are essential on all but the smallest farms. On a large farm under a single management, power needs can be scheduled by careful timing of plantings and by the purposeful selection of crops that have complementary power needs during their growth cycles. On small farms, however, although the power required for intensive multiple cropping is much less, timing is even more crucial. The near-perfect coordination of cropping schedules that can be managed on larger farms is seldom possible on smaller units.

Moreover, the smaller power units used on small farms have inherent disadvantages: they are relatively ineffective in adverse soil conditions and they operate poorly in fields with a lot of stubble or plant refuse. On the other hand, such small power units can often provide important secondary power for trans-portation, water pumping, pest control, harvesting, or thresh-ing, thus influencing the use of other resources, such as labor or water.

It is critical to the success of any intensive cropping scheme that nearly all production recommendations—crop, cropping intensity, row spacing, seedbed configuration, and others—accord with the type of power and its availability. Power is a key determinant of labor productivity, making possible the progress through increasingly advanced development stages illustrated in Figure 1. By itself, of course, power will not improve the productive capability of the land except where labor is the prime limiting factor.

Cash

Cash to pay for such production inputs as seed, pesticides, fertilizers, etc., is another primary determinant of cropping potential in some cases. Vegetable crops, for example, require a great deal of pest management and fertilizer. Unless these necessities are provided, the productivity of vegetable crops will be low. On the other hand, some field legumes, which supply their own nitrogen, require less cash for inputs. On small farms, human labor can be substituted for cash inputs up to a point. Weeding, for example, can be done with commercial herbicides, by expensive machinery—or by hand. Similarly, commercial fertilizer can be supplemented or replaced by nutrient recycling and composting as long as the required labor is available.

Intensive crop sequences almost always require large doses of nutrients. On many tropical soils, a single cereal crop per year may yield acceptably with the addition of only modest amounts of fertilizers, especially if phosphorus is not limiting, permitting the cultivation of lowland rice. But when a second cereal crop, such as maize or sorghum, is added to the system, the total nutrient requirement increases severalfold. With current technology, it is not possible to grow upland cereal crops following lowland rice on the many low fertility soils, even with irrigation, because of the cost of the fertilizer required for the second crop. Various grain legumes have a much greater potential after rice simply because of their lower fertilizer requirements.

In short, for vegetables or field crops, commercial inputs, purchased with cash, are critical to high-intensity cropping unless human labor is extremely plentiful. Where cash is scarce, labor can be substituted to some extent, but it is important to match the cropping pattern to the cash supply. If, on the other hand, it is another resource—markets for produce, or chemical inputs—that is limited, cropping systems must be chosen to make the optimum use of the scarce resource. On soils with low fertility, for example, a system involving a high proportion of perennial crops may be appropriate.

Market availability and subsistence production

Although it is external to the farm proper, and therefore not treated at full length in this analysis, the availability of the market has a direct influence on cropping potential. Market value minus marketing costs defines the farmer's potential earnings from intensified cropping. To be useful to the farmer, the cash market must be accessible, it must be relatively stable, and it must give the farmer sufficient warning of changes in demand and prices to enable him to plan his yearly cropping pattern. In commercialized agriculture, the farmer's ability to use his resources to their fullest potential is dependent on properly functioning markets. National plans for increasing cropping intensity must necessarily include evaluation and perhaps improvement of market systems.

The domestic, non-cash "market" for farm produce becomes of primary importance when production resources are limited and the farm family itself consumes much of the total farm production. The design of cropping systems primarily for family needs, with only small surpluses diverted to the cash market, has been largely neglected by development planners— development efforts have tended to be singleminded in their market orientation. Nevertheless, it hardly seems desirable for farmers with few production resources, low incomes, and many mouths to feed to depend for a large portion of their food needs on an uncertain market that includes service costs far higher than their own return on labor.

On the other hand, however, the poor farmer cannot afford to neglect totally his small cash crop production in favor of his domestic needs. A balance must be maintained. A diverse diet, an essential element of the farmer's well-being, is relatively expensive to the low-income producer if he has to purchase it in the commercial market. The inclusion of sources of diversity in his own cropping system is important, therefore, and the properly designed small farm system can furnish adequate diversity with a minimum reduction in cash-crop production.

Secondary factors

Of the several secondary socioeconomic factors that influence cropping potential through their influence on primary determinants, farm size is the most readily measured. It is also one of the key determinants of a farm's productive potential and, by the same token, of the farmer's welfare. The effects of farm size can be seen most clearly when the farm family furnishes both labor and management services and there is no outside income. This latter condition is important because off-farm income can increase cash and capital resources to equal those of a much larger farm, while off-farm employment can reduce the labor and management services available to the small farm, thus influencing the selection of a cropping system.

In itself, farm size is not a critical determinant of cropping potential. It does, however, influence the quality and quantity of labor available to the farm. More important, it influences the intensity of farm management. Assuming the availability of markets and power, the efficiency with which farm resources are used depends largely on the availability of family labor and management services. The following discussion assumes, then, that the farm family is primarily involved in farm production, for which it provides most of the labor and management, participating in few outside activities.

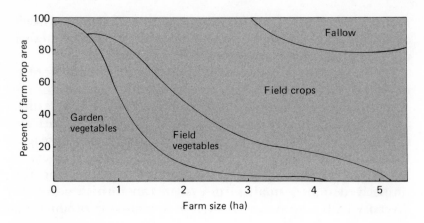

Figure 9.
Relationship between farm size and type of crop in areas of Southeast Asia having a two-crop (eight-month) growing season and access to markets for high-value vegetables. The family is assumed to be the primary source of labor.

On farms of less than 1 hectare, intensive vegetable cultivation, sometimes called "vegetable gardening," is common (Fig. 9). Typical gardening methods include trellising, ditching-and-diking, and intensive intercropping and relay cropping. Some field crops may be grown as well, but on small farms these are usually for consumption by the farm family. Little mechanical power is used. If accessible markets for vegetables are limited, however, more field crops will be grown.

Vegetable production changes with increasing farm size. On farms larger than 1 hectare, field vegetables are grown as the sole crop and power is used for tillage. Field vegetables are seldom intercropped, but they are occasionally planted in relays. A 2-hectare farm may be evenly divided between vegetables and field crops. On farms larger than 3 hectares, a portion of the land may be fallowed for at least one season because of a shortage of power, labor, or management resources.

The labor supply is not often limiting on 2- to 3-hectare farms because hired labor is commonly available and often used. In addition to labor, however, garden fruits and vegetables, and field vegetables as well to some extent, require intensive management. Almost every operation involves judgments and decisions. Vegetable growing is usually diversified in order to use resources efficiently and to achieve some measure of stability; it is also, by the same token, a complex business. Saving or buying seed, planting, controlling pests, harvesting, controlling quality, marketing—all require a kind of management that is not to be had from the hired labor available to small farms. Even a large, extended farm family typical of Asia can supply such intensive management to only 1 or 2 hectares. As farm size increases, therefore, the field crops, which require less management, become more practical and more profitable.

Field crops are better suited to larger farms, too, because they are better adapted to mechanization than are vegetables. A high proportion of the operations required for field crops can be done with animal power or mechanical tillage equipment. The ratio of labor required for primary tillage to labor required for other production operations in a rice-maize-cassava intercrop is very low.

The addition of various power sources as farm size increases, and the effect of power on farm productivity, is shown in Figure 10. The farm sizes to which different power sources apply on typical Asian farms is illustrated in Figure 11.

Where adequate markets for vegetables do not exist, even farms of less than 1 hectare must grow field crops, usually in intensively managed mixtures. If off-farm employment is available as an alternative, however, the small farm may be managed instead like a larger farm. Labor and managerial skills, diverted to income-producing jobs, become scarce factors on the farm; on the other hand, more cash is available for inputs or capital investment in labor substitutes.

The productivity of the land also changes with farm size. Because they are typically more intensively managed, small farms are often more productive than larger ones. Higher productivity can be measured in yields of field crops, in cropping intensities, and in proportions of high-value crops. In Luzon, Philippines, the smaller rice farms have higher yields per hectare than the larger farms, and small upland farms have a higher proportion of their land in vegetables.

The social and economic status of the farmer is occasionally a factor in his willingness to grow unknown, unpopular, or high-risk crops. A high risk of failure may be least acceptable to the farmer whose status in the community is already low. In

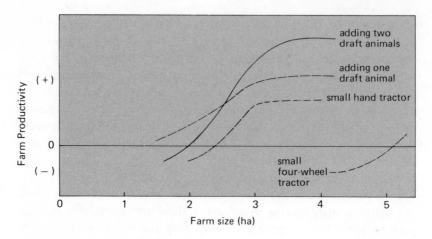

Figure 10.
Change in total farm productivity through addition of various power sources in area where the growing season is six to eight months.

Batangas, Philippines, it is the older and more respected farmers—those with the most to lose, in the commonsense analysis—who often prove most willing to experiment with new crops and technologies.

Another social factor influencing the farmer's agricultural choices is his sense of security against theft. High-value or especially prized food crops are obviously the most in need of protection against theft, increasing the costs of labor and management as well as the farmer's financial risk. For example, crops such as watermelon, sweet maize, and groundnuts often attract thieves. Besides human predators, the farmer may also have to guard his high-value crops against birds, rodents, insects, and even domestic grazing animals.

Land tenure influences the farmer's decisions about whether to make improvements, what cropping pattern to select, what he should do to improve and maintain soil fertility, and what investment he should make in weed control. Some farmers have been known to invest several seasons of labor and management in weed control against nutsedge on newly rented land that they confidently expect to be using for many years.

Farm layout also significantly influences the farmer's ability to manage his crops. If his land is fractioned into a number of scattered parcels, as is often the case, intensive management,

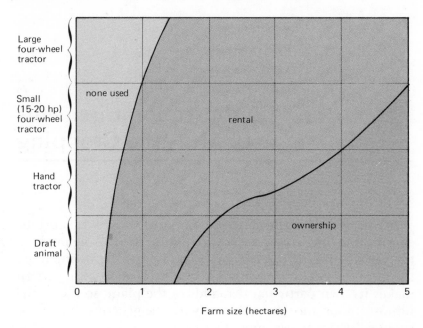

Figure 11.
Schematic relationship between farm size and power source commonly found on developed farms in Southeast Asia.

efficient use of labor, and adequate crop security may be difficult, if not impossible.

The optimum use of his farm resources usually involves the farmer in some degree of diversification. For a single crop, it would be relatively simple to design a plan that would make optimum use of the available physical resources and environmental factors. A combination of several crops, however, entails a combination of different requirements for labor, power, management, and cash flow. For the sake of efficiency, the farmer must select cropping systems that have complementary requirements, achieving the most uniform possible resource demand that approaches as nearly as possible the maximum sustainable level that his resources permit.

A farmer with 2 hectares, for example, cannot plow, plant, weed, or harvest it all at once, especially if he has but one draft animal and little surplus labor. He must therefore plan to stagger his plantings, choose different cropping patterns for different fields, and pursue complementary enterprises as he seeks to use his resources efficiently.

8
Resource Requirements
of Multiple Cropping

In discussing multiple-cropping alternatives it must be emphasized that in any crop combination or growing method, the variety and the appropriate technology for growing it are the most important elements. Each multiple-cropping pattern makes its own particular demands of the plant, so it is of the utmost importance that varieties with the genetic capacity for superior performance are chosen.

Multiple cropping is the growing of more than one crop on a single field in the same year. The term "multiple cropping" has become vague with use because of the many different types of cropping patterns it has been used to describe. Its meaning should not be limited—as it often is—to any one type, such as intercropping.

Following a period of intense specialization in single crops during the 1960s, crop production research has broadened to include all aspects of multiple-cropping systems. With the increasing interest in multiple cropping, the agricultural development specialist must distinguish clearly among the various possible cropping patterns and their specific uses in multicrop combinations. Therefore, this chapter deals separately with the different patterns of annual crops, perennials, and annual-perennial mixtures. Animal husbandry will be considered in the next chapter.

Crop sequencing

Intensified cropping with annual crops usually involves

crop sequencing—planting a second crop after the harvest of the first. The sequence may preclude extensive seedbed preparation between crops. Any number of crops can be grown in sequence, depending on the availability of resources and the local environmental factors. A crop sequence is not as complex as many of the relay or intercropping patterns; it has the greatest potential for increasing productivity on the largest number of farms, and it is the easiest type to implement. Therefore, it should always be considered first in any attempt to intensify cropping.

A change from a single-crop pattern to a sequence of two or more crops also involves several major changes in the use of farm resources. The first is an increased demand for farm management, because a two-crop sequence requires far more precision in both the timing and the performance of farming operations than does a single-crop system. Increasingly complex and critical management decisions about each aspect of the production operation must be made as various physical and economic resources are stretched toward their maximum potential use, leaving little room for errors. Indeed, an intensification from one crop per year to two more than doubles the management required. The farmer's role as a decision maker and manager becomes more important than his role as a laborer.

A single crop grown in a particular way has certain fixed and predictable power, labor, and cash flow requirements. Furthermore, the requirement for labor, for example, will change from each stage of the growing process to the next: from land preparation, to seeding, to weeding, to pest control, and finally to harvest. Researchers call the changing use of labor (usually recorded weekly in terms of man-hours per hectare), the crop's "labor profile." The typical labor profile—as well as the similar power profile—for any cultivated crop will show peak demand for the resource at planting, during early weeding, and at harvest. Obviously, a small farmer will not be able to do all his planting, weeding, or harvesting at the same time; he cannot meet the demand for labor or power. Therefore, he must spread the demand by staggering his plantings.

In an intensive cropping sequence, however, the farmer has

relatively little latitude for changing his planting times unless he changes his crop. A farmer who is growing two successive crops in a six-month period of heavy rainfall does not have many options on the timing of his cropping operations; his choices are limited by the availability of water. Thus, the first rule for intensified cropping is to maintain the greatest possible flexibility in the timing and resource requirements of cropping systems in order to permit the highest possible level of overall farming intensity.

A vivid example of this crop integration occurred as part of the International Rice Research Institute's on-farm research program in Iloilo Province in the Philippines. The cooperating farmer operated 2 hectares. Traditionally, like his neighbors, he grew one crop of rice per yer. In 1975, wanting to take better advantage of his five-month season of rainfall greater than 200 mm per month, he decided to multiple-crop most of his land with a second crop of rice.

The farmer divided his 2 hectares into four parcels of approximately equal size. On Parcel A he prepared the seedbed in May and began transplanting the traditional single crop of rice in June, using an improved, late-maturing, 140-day variety. On Parcel B he began land preparation with the first rain in April, direct-seeding a 95-day variety as soon as the land was ready and moisture was adequate. Parcel C was prepared later but direct-seeded a day or two before Parcel B with a 110-day variety. Parcel D was direct-seeded at about the same time as Parcel A, but with a medium-maturity, 120-day transplanted variety.

For the second rice crop on each parcel, the time required for land preparation was reduced by using transplanted seedlings, which do not require such thorough work in advance or as much time in the field to mature. This sequence, while not optimum, permitted the farmer to make efficient, continuous use of his single draft animal from April to November. Since each crop matured at a different time, he was able to spread out the peak labor requirements for harvesting and planting.

Thus, by changing from single- to double-cropping, the farmer lost some flexibility in his planting schedule, but he preserved his overall advantage by using two planting methods—direct seeding and transplanting—combined with three varieties of rice with different growth periods to spread out the demands on his resources. As a result, the labor and power profiles for his operation were quite flat, with demands on his resources spread fairly evenly throughout the growing season.

Sequential cropping without the use of chemical herbicides to control weeds requires an increase in total power for mechanical tillage. The need for rapid and timely tillage effectively rules out sequential cropping where neither animal nor mechanical power is available. Where human labor is used instead of animal or mechanical power for tillage, the need will be for heavy manual labor rather than the lighter labor characteristic of many other field operations. Sequential cropping generally requires large amounts of heavy labor but small amounts of light labor, or tending, while in common intercropping patterns the proportion of heavy labor to light labor is usually the reverse.

Fertilizer requirements in multiple-cropping systems are determined by the needs of the particular crops, but in general double cropping requires far more than twice as much fertilizer as single cropping, especially if both crops are cereals. In many soils of the humid tropics, normal nutrient cycling permits a single crop per year with only modest applications of fertilizer. Under most conditions, for example, a single crop of lowland rice will produce an economically optimum yield with about 40 kilograms of nitrogen per hectare. To double this yield by growing two crops, however, will require three to four times as much nitrogen. The fertilizer requirement can be reduced by substituting a legume for one of the cereals in the rotation.

Nutrient requirements are especially critical when upland crops are grown following lowland rice. The submergence of the soil creates a chemical environment in which phosphorus is readily available to the plants. When the field is drained and the soil dries out for the upland crop, supplies of phosphorus and the organic matter that decays to provide nitrogen are reduced. Thus, lowland rice can do relatively well in conditions of low soil fertility that are submarginal for following upland crops. At the Chainaut research station in the central plain of

Thailand, for instance, it was found that upland cereal crops such as maize and sorghum were not economical following lowland rice because of their high fertilizer requirements. The legume mung bean, which provides its own nitrogen supply, proved to be the most economical alternative.

Tillage is often a problem in intensive cropping sequences. To fit two or more crops into the period when adequate water is available, the farmer may be forced to begin land preparation before soil moisture is at the proper level. Tillage under these conditions is expensive and difficult with animal power or small machinery. On the other hand, the heavy soils typical of rice-growing areas in the humid tropics are difficult or impossible to till when the rainfall exceeds 250 mm per month. If the second crop is an upland crop, the combination of high rainfall and heavy soil may prohibit tillage during the brief turnaround time between crops.

An even more difficult tillage problem involves converting the soil from the puddled condition necessary for lowland rice to the granular condition required by upland crops. Legumes can be planted successfully with several low-tillage methods which usually involve a light harrowing to loosen the surface soil. Vine crops, such as watermelon, can be planted after the puddled soil has dried with no more tillage than the digging of small holes, widely spaced, for the transplanting of seedlings. Soybeans are commonly planted with "zero" tillage in Taiwan and Indonesia. The seed is sown in holes dibbled in the rice hills from the previous crop. This method works only in certain types of soil.

In general, upland crops following paddy rice on heavy soils are planted at the end of the rainy season, when they will face both poor soil conditions and moisture stress. Low-lying, well-drained soils with relatively high water tables are therefore especially well suited to upland crops following a main rice crop. Freshwater alluvial deposits are thus generally well suited to multiple cropping while marine deposits are not.

Where irrigation is available, puddled soil can be converted to upland conditions with rotary tillage as the soil dries, followed by one or two cycles of wetting and drying to fracture clumps of soil. This expensive process of swelling and shrinking the soil clods is suitable for heavier soils that have a high montmorillonite clay content where there is less than 50 mm of rain per month and the paddy can be drained and dried.

Sandy soil types can be converted from puddled to upland condition by plowing before they are completely dry. But most rice soils are difficult to convert successfully from puddled to upland conditions, especially under uncertain rainfall conditions. Those soils are best adapted to double-cropped rice.

Crop stubble and residue can interfere with rapid tillage operations during intensive cropping sequences. Animal-drawn implements and small machines can handle only limited amounts of stubble and residue. The problem is especially significant when the crop turnaround takes place in wet weather, when the traditional technique of burning off crop residues becomes impossible. Since the removal of crop refuse from the field requires a lot of labor, second crops are commonly planted with minimum tillage in the stubble of the previous crop. In any event, the effects of crop stubble on tillage must be carefully considered in the planning of intensive sequences.

Still another important factor in intensive sequences is the crop's growing period to maturity. The relatively recent increase in interest in double-cropping rainfed rice, for example, is a consequence of the availability of improved, fast-maturing varieties from IRRI. Unfortunately, however, in general not enough attention has been paid by plant breeders to early maturity in maize, groundnut, and rice. As a result, even the earliest-maturing improved varieties are still considerably later than the earliest traditional varieties. A wide range of maturities is essential for the planning of intensive multicrop sequences.

In crop sequences, the problem of weed control takes on a new aspect. In a single-crop pattern, weed populations are usually determined by their growth and reproduction during the long fallow period. But as cropping is intensified and the land is either being tilled or cropped during most of the period when moisture is available for plant growth, weeds adapt to the new conditions. In the typical sequence of upland rice and maize, the tall, leafy rice varieties form a dense canopy after about 50 days of growth, while the maize is planted densely in closely spaced rows. Thus, both crops are effective in shading out the light-sensitive sedges and grasses. The broadleafed weeds are more effectively controlled by frequent tillage, using animal power. After the maize harvest, the land is plowed to establish a clean fallow for the dry season. If necessary to control nutsedge, the field will be plowed several times during the dry season to expose nutsedge tubers to dessication.

In experimental plantings of multicrop sequences, it is common for weed populations to shift toward troublesome sedges and grasses. In order to maintain the economic viability of the system, it is important that this shift in weed growth be controlled by adjustments in planting, tillage, and weed management. Judicious use of herbicides can be extremely effective, but chemicals should never be the sole means of weed control. The maintenance of high herbicide levels will eventually encourage a shift toward the more resistant and difficult to manage weed species, resulting in a more severe weed problem.

Insect problems in multicrop sequences, on the other hand, are seldom more serious or complex than they are in single-crop patterns. In lowland rice culture, the continuous cropping of irrigated lands may contribute to the buildup of plant viruses and their insect vectors; such buildups have not been found, however, in upland areas or humid areas where upland-lowland crop sequences are grown.

One exception to this rule is the European maize borer, a pest on cotton and sorghum as well as maize. Because it is especially vulnerable to this pest, cotton does not perform well in rotation with maize or sorghum. In addition, in upland areas that are intensively cropped, nematodes are potentially an even more serious pest, although they have not proved to be a problem yet in the high-rainfall areas that have been surveyed. Nematodes have become a problem, however, for intensive crop sequences that include a highly sensitive crop such as tomatoes.

Disease buildup has occurred in intensive sequences of continuous rice and double-cropped maize. In humid areas, downy mildew is a serious disease of maize, making it necessary to rotate other crops with maize in the Philippines, Thailand,

Indonesia, and other countries. While few diseases affect more than one crop, some soil-borne pathogens seriously affect several species of legumes. Research in the Philippines indicates that the increase of soil-borne pathogens causes a buildup of soil toxins that reduce the growth of future legume crops. Therefore, although they fit well in multicrop sequences, legumes should be planted continuously or in intensive sequences only with great caution.

Relay planting

Small farmers occasionally practice relay planting—sowing a second crop in the sequence after the flowering of a first crop, but before its harvest. In Southeast Asia, for example, the common relay-planting practice is to seed a grain legume in a field of standing rice during the last week before the rice harvest. The flooded field is drained and the grain legume seed is broadcast onto the moist soil beneath the rice plants.

The purpose of most relay planting is to save time when it may be difficult to fit two complete crop cycles into the growing season because of temperature or water limitations. The time saving may be greater than just the few days during which the two crops overlap: if the second crop is not seeded before the harvest of the first crop has begun, labor for planting may not be available until the harvest is completed. When zero-tillage techniques are used, relaying has the additional advantage of establishing the second-crop seedlings in the shade of the first crop, whose canopy maintains moisture at the soil surface.

Where labor is plentiful and a high level of management can be sustained, relay planting is sometimes practiced on farms of 2 hectares or less. In Taiwan, rice may be relayed with sweet potato, which requires a relatively coarse-textured soil. Or, bean or cassava may be relayed with maize. Relay planting is a common technique for intensive vegetable production on small farms.

A number of factors add to the difficulties of relay planting. Good weed control during the first crop is a prerequisite for the success of the second. Moreover, the seeding of a second crop into a standing crop usually involves more labor and more difficulty than open-field planting. Seeding with low or zero tillage increases the risk for the second crop, and if conditions are not just right the resulting yield can be severely affected.

For all these reasons, relay planting should only be attempted when there is a significant benefit to be gained and when conditions appear favorable. The possibility of saving a few days does not often justify a significant sacrifice in yield. While relay planting at its best is highly productive in terms of yield per unit of land, its return on labor and management may be less generous because it requires so much of both resources.

Relayed crops differ in their tolerance of the shade of the standing crop. For crops relayed into standing rice, the maximum overlap periods are:

mung bean	2-3 days
radish	2-3 days
maize	1 week
soybean	1 week
sorghum	2 weeks
sweet potato	4-5 weeks
cassava	several weeks
taro	several weeks

Longer overlap periods will severely reduce the yield of the relayed crop.

Fertilizer is usually applied to the relayed crop after the first crop has been harvested. The first and second crops in a relay system are always different, to reduce the opportunity for pests and diseases to carry over. A potential difficulty in relay cropping is the lodging of the first crop, making the seeding of the second crop extremely difficult and slow. The farmer may also experience difficulties in using power equipment for seedbed preparation, in controlling weeds, and in managing the seedling stage during the early days of the second crop before the first crop has been removed.

The potential of relay cropping is enhanced if the farmer splits the harvest of the first crop, harvesting alternate rows earlier than the rest. This system is practicable with crops that

A complex intercropping of cucumbers, beans, celery, and chives in China.

can be harvested either green or dry, such as soybeans or maize. The second crop will also benefit if the farmer strips excess leaves from the first crop. Used with care and attention to details under the proper conditions, relay planting can greatly increase productivity. Used indiscriminately, however, it will not be productive.

Intercropping

Intercropping of annual crops—planting two different crops together in the same field at the same time—is the least understood of all the cropping methods. Until recently little research of any kind had been done on intercropping patterns; even now the complexity of many intercropping mixtures discourages attempts to understand them, and most of the attention is drawn to the simpler examples.

A few generalizations can be made about intercropping. First, each of the many possible intercropping patterns is appropriate for a particular situation or range of conditions

and inappropriate for others. Second, a particular intercropping pattern is almost always chosen to alleviate a
particular limitation in resources. Third, intercropping is
almost always associated with farms of less than 2 hectares,
usually of less than 1 hectare. Fourth, any intercropping
pattern must be designed with careful attention to the details of
plant type, planting arrangements, timing and other factors.
Fifth, intercropping combinations make it difficult if not
impossible to cultivate between the rows with animal- or
tractor-drawn equipment.

There are several types of intercropping mixtures using
annual crops. The most common type is a mixture of short and
tall plant types in which both crops are planted at the same
time, but the taller crop is harvested first. Maize might be the
tall crop, harvested after three months, for example, with the
intercrop of groundnut, sweet potato, or rice harvested after
four months. Such intercrops of different plant types with little
competition between crops during the reproductive stage are
usually the most productive.

Another pattern mixes two tall crops with different growth
rates or different planting dates so that one matures before the
other. Three-month maize or sorghum, for example, can be
intercropped with cassava, which takes 10 months to mature, or
pigeon pea, which takes 8 to 10 months. Such intercrops are
usually more productive than monocultures.

A third type of intercropping involves a crop of short
duration and short stature, like soybean or a vegetable, planted
at the same time as a taller, slow-growing, long-maturing crop
such as sugarcane. It is especially important that the two crops
complement each other. In the case of sugarcane, for example, a
variety that is slow to close will be helpful to the intercrop.

Sometimes an early-maturing, shorter plant is intercropped
under a taller plant—mung beans under maize, for example—
but the productivity of the combination is uncertain although
there may be compensating benefits.

A fifth type occurs in Africa, where farmers commonly
intercrop plants of similar type but different maturing periods:
millet or maize, which mature in three months, for example,
with sorghum, which matures in six months.

An Indonesian farmer and his field of upland rice, maize, and cassava.

Intercropping can be used to achieve a number of different agricultural objectives; for each objective, certain specific intercrop combinations are most appropriate:

• Higher overall productivity can often be achieved with crop combinations such as maize-upland rice, in which the plant types are different and the growing periods do not completely overlap. The two crops are planted at the same time, with the maize at a density of 20,000 to 30,000 plants per hectare in rows 2 to 3 meters apart. Rice is drilled between the maize rows in rows 25 cm apart.

In this maize-rice intercrop, the growth patterns of the two crops are mutually complementary. The maize grows more rapidly than the rice, accumulating dry matter at a high rate during the first two months, and is harvested before the heading of the rice plants. The rice, its growth only slightly retarded by the maize, then proceeds to maturity with a relatively high yield. The difference in the maturing periods of the two crops is the critical factor: ideally, the maize is harvested as dry grain 80 days after seeding, while the rice is harvested 125 days after seeding. In research plantings, this intercrop combination has yielded 60 percent better than monocultures. Other combinations showing a similar yield advantage include maize-cassava, maize-beans, maize-groundnut, and sorghum-millet. Such long-duration but slow-starting crops as sugarcane and pigeon pea can be profitably intercropped with a fast-growing crop that can be harvested early.

• Other combinations may have higher labor productivity, especially the long-duration combinations used when power for tillage is not available. If the soil must be tilled by hand, the combination of short- and long-duration crops results in a long cropping season with multiple harvests, giving the maximum return on the labor invested in the initial hand tillage. In a maize-cassava combination, for example, the cassava covers the soil after the maize is harvested until it is ready for harvest itself in 10 to 11 months. The sorghum-pigeon pea combination offers a similar advantage.

These combinations may be planted on part of a farm only, with the rest planted in a two-crop sequence. During the midseason, when labor for harvesting and replanting the sequence is scarce, the intercrop requires little labor. Many combinations offer the advantage of a high ratio of light to heavy labor, and are therefore well suited to small farms where the farm family does all the work.

• Once they have formed their canopy, long-duration intercrops generally do not require tillage during their growing period. Thus they can tolerate wet periods when tillage is difficult or impossible.

• Weed control is simplified in many intercrop combinations because the rapid establishment of a dense canopy reduces weed growth. Shade-sensitive weeds such as nutsedge and *Imperata cylindrica* may be eliminated entirely by a combination like maize-mung, which intercepts 90 percent of the incident light after 50 days of growth. Maize alone intercepts only 80 percent of the light. Continuous high-density intercropping will eventually eliminate all light-sensitive weeds from the field.

• In certain instances, intercropping has been found to improve control of insect pests and diseases. In Southeast Asia, for example, it has been shown that maize in rows 2 to 3 meters apart, intercropped with soybean, groundnut, upland rice, or mung bean, suffers relatively little from downy mildew, normally a major maize disease. The wide spacing of the maize rows also reduces the incidence of the maize borer, a major pest of maize. When maize is intercropped with groundnut, the number of maize borer pupae is reduced as much as tenfold. While few other examples of insect or disease advantages from intercropping have been found to date, future research may find other useful crop combinations.

• There is some indication that long-duration intercrop combinations have an advantage when nutrients are limited. The lower populations of the longer-duration plants take up nutrients at a slower rate than a high-density population of a short-duration crop. In contrast, high-density, intensely managed monocultures of maize or rice are efficient in taking up applied nitrogen, and in situations of sufficient nitrogen they recover rapidly from nutrient stress and yield well. However, they grow poorly when nutrients are limited and only slowly released.

• A great deal has been written about the inherent stability of crop mixtures—the so-called "insurance" factor. There is evidence to indicate that when one of the crops in a combination is damaged early in the growing season by adverse agroclimatic conditions, pests, or diseases, the other crops may compensate for the loss by doing better. There is little available data to indicate how general or important this phenomenon is in actual practice. Nevertheless, increased yield stability is commonly accepted as an advantage of the complex, multicrop mixtures used by farmers who practice shifting cultivation, and a similar advantage may well accrue to farmers who intercrop as well.

In summary: mixtures of annual crops are used by farmers for a variety of specific reasons; different mixtures have different characteristics that give them specific advantages in particular situations; therefore, care must be taken not to generalize about intercropping without being specific as to the type of intercrop involved and the situation in which it was applied.

Perennial crops

Perennial crops are important in mixed farming systems. Relatively little needs to be said about them here, however; instead, the reader is referred to the excellent review by H. Ruthenberg (see Bibliography).

Perennial crops—shrubs and trees—remain a grossly under-exploited potential resource in small farm agriculture. Many perennials are especially well suited to marginal lands with steep slopes or inherently low fertility. Such perennials as coconut, cacao, and a great many tropical fruits and nuts should usually be grown by the small farmer in mixed plantings, often widely spaced to allow staple food crops to be planted between the rows. As cash crops, perennials should not be grown as a sole crop by a small farmer who does not have enough additional land and labor to produce food crops to satisfy at least a large part of his domestic needs. The farmer with few resources can ill afford the uncertainties and the high service costs involved in purchasing most or all of his food in the retail marketplace.

Most perennial crops produced by hand labor afford the small farmer few economies of scale. Therefore, the small farmer might well begin his plantation with a small number of young perennial trees or shrubs interplanted with his regular annual crop. When annuals and perennials are cropped simultaneously, the added cost of maintaining the perennials during their early, nonproductive years is low. The particular mix of crops, and the eventual balance between annuals and perennials, should be decided according to the local market for the cash crop, the type and size of the farmer's land, and the food needs of his family.

In parts of Asia where such plantings are well developed, perennial crop mixtures in the noncommercial homestead area around the house and the farmyard contribute greatly to family comfort and welfare. Farmers in Indonesia have considerable experience with a three-tiered crop canopy in which a mixture of crop species provides shade, privacy, dietary variety, fuel, building materials, and a cash crop as well. The system takes advantage of the nutrients from animal, plant, and human wastes that accumulate in the farmyard. The development of such a system is more likely in an extended village than in a clustered community because the extended pattern affords more space for crops around the houses, where wastes can be accumulated and concentrated. In systems that allow animals to graze freely, potential crop nutrients are dispersed, and homestead crops are more difficult to establish and less productive.

As the attention of development planners focuses on the total welfare of the farm family, rather than on the yields of major cereal commodities, the potential importance of perennial crops in mixed farming systems deserves careful assessment. Perennials offer many advantages in multicrop systems. On low-fertility soils where annual crops would soon exhaust the available nutrients without substantial fertilization, perennials can sustain a high level of continuous productivity. Over a number of years, trees accumulate large quantities of nutrients, and their extensive root systems are able to take advantage of soil moisture and nutrients. Trees also recycle nutrients in a closed system, dropping leaves and twigs and then reabsorbing their decay products from the upper soil layers. Moreover, whereas an annual cereal plant typically devotes 25 percent of the nutrients it absorbs to the formation of grain, a tree uses only a small fraction of its nutrient supply to produce a heavy crop of fruit or nuts. As a

Table 2.
Types of cropping patterns commonly used in Asia, their energy requirements and productivity characteristics.

	Energy input			Productivity
	Human	Power	Chemical	
Slash and burn cultivation (intensive mixtures of crops)	H	0	0	VL
Tree monoculture	L	L-0	L-0	M
Trees with annuals intercropped	M-H	L	L	H
Single crop monoculture	L	L	L	M
Multicrop sequences	H	VH	VH	VH
Intensive vegetable production	VH	L	VH	VH
Intercropping of annuals	H	M	M	H

Key: VH = very high; H = high; M = medium; L = low; VL = very low; 0 = zero.

result, the coconut tree, for example, can produce a substantial crop year after year on soil too infertile to support a single good maize crop.

In addition, perennial crops are stable producers. They are not seriously affected by short-term fluctuations in rainfall, and they contribute to the overall stability of the cropping system by not requiring precise or timely field operations. They require only low levels of labor and power.

Table 2 summarizes the resource requirements and productivity of various types of cropping patterns. The critical factor in choosing among these alternative systems is their resource requirements. Where resources permit, the most productive system is the best choice. However, systems with lower resource requirements can also be highly productive. They should be carefully considered in development efforts aimed at farmers faced with severe resource constraints.

Animals in Mixed Farming Systems

Raising livestock and small animals in the tropics is an industry divided into two distinct sectors. Large, commercial enterprises, especially those producing milk, swine, and poultry, are found throughout the tropics, but they represent a large portion of the total animal industry only in some parts of South America and a few other areas.

These large animal enterprises are generally identical in organization and methods to their counterparts in the developed countries. With large capital investments and a high degree of specialization, many of these commercial animal farms do not even grow their own feed.

Not surprisingly, this commercial sector of the animal industry is the focus of most research and development efforts. As a result, a great deal of knowledge has been developed about modern technology for large-scale animal raising and the organization of this technology into ideal production systems for each animal. These technologies and systems have been locally adapted for use in each area.

Contribution of animals to mixed systems

The second sector of animal enterprise, combining animals with mixed cropping systems, is far more prevalent in most developing countries. Few small farms in the developing world are without animals. A common characteristic of these small farms with mixed animal-crop enterprises is the close interaction between animals and crops. The complexity and variety of these interactions have discouraged research and development efforts from being directed at these ubiquitous, very important mixtures. Nevertheless, the inherent efficiency and productivity of these animal-crop combinations, due to the interactions they permit, are responsible for their almost universal popularity among farmers.

The efficiency of crop-animal interactions is most pronounced where production resources are scarce; therefore, they are crucial to the improvement and success of small farms, where production potential is otherwise limited, and where the possibility of capturing and exploiting additional crop and labor energy by raising animals is of relatively large importance. In the most productive crop-animal interactions, the animal is used as a source of power to farm the crop, as a source of meat for domestic consumption and for sale, as a consumer of crop byproducts, and as a means of recycling nutients into crop land (see also Chapters 2 and 11). This near-total interaction is the most common animal-crop relationship on small farms in Asia, and its productivity accounts for the increasing popularity of this type of system in that region. Work animals tend to be more common on small farms. They require little capital investment from the small farmer, who can breed them himself or buy them cheaply when they are young. The operating costs of a work animal are also relatively small, making it admirably practical for a farming system in which cash flow is extremely limited.

In areas where there is a cash market for meat, the animals increase in potential value as they grow. The better draft animals, however, are usually not sold until they have outgrown their value as workers, when they bring reduced prices at market. The raising and maintenance of draft animals is usually combined with the raising of other animals purely for meat. Both types of animals require tending rather than heavy manual labor, and the work is usually assigned to otherwise unemployed children or old people.

Sources of feed

The two principal ways of feeding animals in mixed farming systems are controlled grazing and cut-and-carry feeding.

Controlled grazing depends on the availability of suitable pasture land, either unused farm land or commercial grazing areas. To some extent, animals can also graze successfully on the stubble that remains after crops have been harvested, along fence rows, or among trees planted on the farm. As farming is intensified, bringing each piece of land to its most productive use, the need to harvest feed and carry it to the animals increases. Often a combination of controlled grazing and cut-and-carry feeding is the most productive relationship between the crop and animal systems.

On smaller, more intensely worked farms, animals are seldom allowed to graze freely. Instead, feed is harvested and carried to the animals. Where intensive cropping is interrupted by a period of little or no rainfall, feed is stored at the end of the wet season to carry the animals through the dry months. In Batangas, Philippines, an area of intense cropping and close interaction between animals and crops in the typical farming system, a large portion of the land is planted in maize late in the wet season. The planting is timed to bring the maize to maturity after the rains have ended. The lack of moisture late in the growing season often reduces grain yields, but the loss is tolerable as the price of providing stover that can be dried in the field, carried to the farmyard area for storage, and fed to the animals throughout the dry season.

In many cultures, communal grazing areas are a traditional part of the village. Common grazing makes disease control and selective breeding almost impossible, however, and the productivity of animals maintained in this manner is often low. As farming becomes more intensive, especially in areas of high rainfall, less land is available to provide off-farm sources of feed. In many areas in the tropics, land use has already become so intense that almost all fodder must come from the farm itself.

If the farmer must depend on his own land to provide most of the forage for his animals, he can make several adjustments in his cropping patterns. His approach will depend to some extent on the physical layout of his farm and the village, and his feeding problems will be far more complex if his house is at some distance from his farm land. He may take his animal feed from the fence rows or from areas planted in tree crops, where he can also grow a species of saccharum that is competitive with other grasses and weeds and highly palatable to animals, but that will not invade cultivated fields.

The lengths to which a farmer will go to get animal feed, and especially the extent to which he will devote crop land to that purpose, depend on the size of his farm and the scarcity of food. If his farm is less than 2 hectares, and he values his animals highly for both work and meat, he may plant some of his land in feed crops. In rice-growing regions where carabao are used as draft animals, the rice straw is used as a supplemental feed, easing the need to devote crop land to growing sustenance for animals. Cattle, however, require higher quality feed.

In Batangas, Philippines, feed is in short supply during the dry season, as well as in the early part of the wet season, when all available land is planted in crops. Because of the scarcity of land, the animals cannot be left to graze. To provide feed, the farmers interplant maize into upland rice at the onset of the monsoon rains, harvesting it 30 to 60 days later as green fodder. A considerable volume of maize is taken from the rice fields without seriously affecting the rice yield. Feed maize is also often planted in dense populations early in the season along the edges of the fields.

There are many crops that can be thinned or pruned to produce animal feed as a byproduct of their regular production. When cereal or vegetable crops are grown in a second season, the plants are often sown closely within the row, then thinned as they grow to improve the stand and also to provide feed. Commercial maize crops are often thinned in the row, stripped of their lower leaves as the grain is forming, and topped as the grain matures, to provide feed. After the harvest, the stalks are cut and stacked for feed during the dry season. Sorghum is thinned and pruned in the same way, especially in India, where it is a common feed crop on small farms in drier areas, and where special sorghum varieties are grown for fodder. Sweet potato vines are also pruned for feed.

Weeds can also be harvested for feed during the cropping season. After the crop harvest, many crops, including most legumes, can be fed to animals in a mixture with other feeds. The limiting factor is often the high fiber content of the mature crop.

Grain is generally fed only to chickens on small farms, in part because the small farm lacks the facilities to grind grain for feeding to other animals. It is almost always more profitable for the farmer to convert his grain into chicken meat than to sell it directly.

The relationship between animals and farm size is conditioned by several factors: the length of the growing season, the amount of rainfall, the overall productivity of the farm, the importance of animals in the cropping system, and the values of alternative crops. In areas like many of the humid regions of Asia, with relatively high soil fertility, a growing season long enough to permit two crops, and high rainfall, one or two cows can be maintained in combination with crops on a farm as small as 1.5 hectares. As the farm's productivity decreases, the amount of land needed per animal increases.

While larger animals such as horses, cattle, goats, and sheep must be fed year-round, the numbers of smaller animals can be changed according to the crop season and the availability of feed. Ducks, for example, grow to marketable size in only a few weeks, so they are well adapted to be raised as a seasonal crop. In Thailand, the numbers of ducks are increased markedly during and immediately after the rice harvest, when the rice stubble and its gleanings are available for feed. The increased supply of ducks reaches the market just in time for the Chinese New Year. Chickens, rabbits, and geese are almost as versatile in their ability to take advantage of seasonal crop residues. An additional advantage for the farmer in raising these small animals is that they can be sold or traded in small transactions, in units of a single animal, as "small change." By the same token, they are well suited to the farm family's dinner table.

Animal management

The management of animals is critical to a productive crop-animal relationship. Where forage animals, especially pigs, are allowed to wander unattended, the productivity of crops suffers significantly. In such villages, crop diversity will be much less than in villages where animals are closely tended and managed. In villages of Mindanao in the Philippines, or in northern Thailand, where pigs and other animals wander

In villages where farm animals roam freely, they eat all plants except those that are unpalatable to man. Thus the family's potential food sources are greatly reduced.

unattended, a farm will have no more than 8 or 10 different food crops. In Indonesian villages, in the hills of Nepal, or in parts of the Philippines, under similar climatic conditions, animals are confined and 50 to 60 economic plant species are commonly found on a single farm. The crops include a wide range of edible plants: root crops, tree fruits, legumes, and cereal crops. This direct correlation between the level of animal management and crop diversity and productivity is found wherever animals and crops are grown together (see Table 1 in chapter 2).

Animal confinement and management therefore often become social issues. The pasturing or grazing practices of one farmer may be at the expense of other farmers in the area as his animals wander onto their land or use more than his fair share of common land. As cropping is intensified and former off-season pastures are given over to second crops, other problems can arise. Farmers often have to fence their land against trespass by their neighbors' animals, at great cost in time, labor, and materials. Efforts to intensify cropping often depend on the passage and enforcement of local ordinances requiring the confinement or close management of animals.

The labor force for such close management is normally made up of the younger members of farm families. When a farmer gets older and no longer has young children at home to tend his animals, he will sell them. Farmers who have more animals than they have children to manage them are often hard hit by requirements for confining and managing their livestock instead of letting them graze freely throughout the village. Such farmers can seldom afford off-farm labor to manage their animals. Not infrequently, these farmers will use their considerable influence in village politics to forestall regulations forcing them to manage their animals more carefully, so changing traditional ways in order to foster a more productive relationship between animals and crops is often a slow, frustrating process.

Advantages of mixed systems

The many benefits of mixed animal-crop systems account for their popularity among small farmers. The systems achieve their highest productivity by using animals to consume crop residues or less popular crops that the farmer could not otherwise market commercially. Animals also provide a valuable addition to the farm family's own diet. Where commercial markets exist, animals represent an extremely valuable source of capital gain and cash income. And they tend to stabilize farm productivity during short-term climatic fluctuations, which have little effect on them.

Where reliable markets are accessible, the farmer can use his animals as a hedge against sudden and unexpected cash needs, selling an animal or several animals to meet the emergency expenses of sickness or death, or of marriage and other celebrations. In the Philippines, it is common to sell a pig to pay school expenses. A recent publication from IRRI notes: "A study of weekly cash flows showed critical financial linkages among the farmer's crop, his livestock, and his household needs. For example, farmers often sell livestock at planting time to purchase needed inputs."

In areas of small farms, then, it is of the utmost importance that both animal scientists and crop specialists understand the interactions between crops and animals and their potential for increasing small farm productivity. In most countries there is little contact between these disciplines; both should be reoriented toward studying the potential of cropping systems to provide feed for complementary animal systems.

Despite the almost universal interest of farmers in mixed crop-animal systems, professionals in both crop and animal production commonly pursue research in pure crop systems or pure animal systems, without reference to the interactions between the two that increase the productivity of both. Fortunately, most farmers have no such inhibitions or prejudices. Science should do more for them. The exploitation of mixed crop-animal production systems deserves the attention and commitment of development policymakers and administrators, as well as of scientists, especially as agricultural production resources become strained to their limits.

10

Noncommercial Farm Enterprises

Noncommercial farm enterprises include all agricultural activities whose products are consumed primarily by the farm family itself. The category also includes, however, those activities that result in small cash sales or trades that are incidental to the farm's principal commercial enterprises.

The farmyard as a center of production

The most important noncommercial activities are carried on close to the farmer's house. This area has different names in different countries: in Indonesia it is called the *pekarangen*; in the United States it is the farmyard. The potential for development of noncommercial enterprises in the farmyard is largely determined by the physical layout of the village. If farmhouses are clustered close together, there is little room for development near each house. Clustering is commonly either a security measure designed to protect lives and property in areas vulnerable to intrusion, or an adaptation to topography, as for example a village clustered on higher ground in a lowland flood plain. The development of farmyard enterprises is encouraged, on the other hand, by village layouts like the Indonesian transmigration settlements, in which houses are purposely spaced along the roads leading to the village centers. Each house has direct access to its own farmland, and development of the farmyard area is unhindered.

The most common farmyard plantings are tree crops. In areas with high rainfall (more than 1200 mm per year) and fewer than six dry months, the predominant tree crop is coconut. Other tall fruit trees that may be grown include mango, jackfruit, breadfruit, rambutan, lichee, and kapok. Leguminous trees such as *Gliricidia* or *Leuceana* may be included as sources of animal feed and firewood, as well as suppliers of soil nitrogen. One or more clusters of bamboo may also be grown for sale or for use in construction on the farm.

Beneath the taller trees, plants of intermediate height, such as coffee, banana, and papaya, are commonly grown. Underneath this second layer of vegetation there may be a third, still lower layer of various shade-tolerant species: ginger, cassava, pineapple, taro, winged bean, and many others. A well-developed farmyard planting essentially mimics the tropical forest ecosystem, replacing the native plant types with economically useful species. Development of the three-layer system of farmyard planting is most advanced on the Indonesian island of Java, where farmers have become expert in selecting and managing the most appropriate species.

Farmyard plantings serve a number of important functions. First, properly selected and spaced, trees provide both shade from the sun and shelter from winds and storms. In typhoon areas this protection is especially important. Second, to the total well-being of the farm family, a well-developed farmyard planting contributes sensory, aesthetic values that, however intangible, may be especially important to low-income, rural people who lack access to more sophisticated cultural artifacts. Third, dense tree plantings afford the farm family a measure of privacy from nearby houses or roads. The privacy afforded by trees, fences, and hedgerows is sometimes socially acceptable in cultures that frown on other forms of privacy. Fourth, noncommercial crops break the monotony of farm diets based on one or two bland staples, while providing vitamins, such as vitamin A, that are not available in sufficient quantities in rice, maize, or other staples. The farm family in the tropics anticipates the seasonal availability of a wide range of fruits and vegetables just as much as more affluent families in the temperate zones; the difference is that the low-income farmer cannot afford to buy his diverse luxuries on the retail market, but must grow them himself. Moreover, the herbs, spices, and other minor food ingredients grown by the farm family at home add considerably to the palatability of many traditional dishes. Fifth, some of the farmyard production may occasionally be sold in the local market, providing the farmer with supplementary cash. Sixth, some farmyard species provide fuel for cooking and heating, and building materials for construction and maintenance. Finally, farmyard crops make effective use of the nutrient sources, such as manure and plant residues, that accumulate near the farmhouse.

In the drier areas, the taller trees in a farmyard planting may be most valuable for shade and firewood. Species are selected specifically for their ability to retain their leaves during the long, hot, dry season. Few species of economic crops can tolerate these stresses.

Several factors influence the development of the farmyard area:

- The physical orientation of the village and the amount of land available around the farmhouse
- The permanence of the settlement
- The climate
- The size of the farm, which determines the amount of land that can be allowed for noncommercial enterprises
- The availability of appropriate species and a knowledge of their uses
- The kinds of animals in the farming system and the level of animal management

Where foraging animals are left untended, the potential for farmyard crops is drastically reduced. In India, where cattle wander untended, and in tribal villages where pigs scavenge freely, farmyards are desolate, with few crop or plant species and few shade trees. The quality of life of the farmers is severely depressed by the lack of farmyard plantings.

Fencerows

Fencerows are often used for noncommercial plantings as well as for their primary functions as field boundaries,

A Nepalese girl carries plant materials for animal feed and compost from pasture to the home garden. The cycling of nutrients from grazing areas to crop fields is essential to productivity of crops in Nepal's intensive hill agriculture.

enclosures for containment or exclusion of grazing animals, and erosion controls. In many Asian areas of high rainfall, long-established fencerows that are highly effective in holding topsoil on terraces are often planted with intermediate-height trees or shrubs such as bananas or papaya. Kapok is also frequently planted in fencerows because it casts relatively little shade on the crops. Leguminous trees such as *Gliricidia* are used in Indonesian fencerows for firewood, as well as for a source of nitrogen for the small fields. Its erect branches create little shade in the fields. Fencerows are also often planted with grasses that are very palatable to animals, but that do not spread to the fields to compete with the crops.

Fencerows may serve as breeding places for pests and diseases in isolated cases, but this is not general. There is evidence to indicate that the plant diversity and permanence of the fencerow make it a refuge for beneficial insects and predators. The relative rarity of pest outbreaks in highly diversified small farm areas where hedgerows and farmyard plantings are extensively used may be due to the net benefits of these traditional features.

Trees scattered in grazing areas or in commercial crop fields are common in much of the tropics. The trees in grazing areas where the climate is unusually hot are valuable for the shade they afford the animals. In crop areas, however, the economic utility of trees depends on the cropping intensity of the field and its productivity. If cropping intensity and productivity are low, tree crops can stabilize or increase the total yield of the field. If intense cropping includes the extensive use of expensive chemical inputs, and the conditions for crop growth are generally favorable, trees should probably be removed to make room for more productive, highly managed crops. Where conditions for crops are poorer, and crop management may be less intense because inputs are unavailable, however, trees should probably be left in place or even planted.

Noncommercial plantings can also make use of gullies and hillsides adjacent to crop areas. Where farms are small, these waste areas are often converted to economic use with intensive, uncultivated plantings of fruit trees or species valuable for firewood.

The establishment of such productive, noncommercial enterprises can add greatly to the welfare of farm families and the productivity of small farms. Nevertheless, agricultural scientists are largely uninformed about such possibilities, or even unaware of the existence of farmyard plantings through-out the tropics. Market-oriented research, concentrating on per-hectare yields of rice, maize, and other commercial crops, overlooks this critical, traditional element of rural well-being. In every country, research on farmyard plantings should develop and disseminate information on appropriate non-commerical crops and their management. This information, and actual planting materials, should be made widely available to farmers.

11
Nutrient Needs of
Intensive Cropping Systems

The achievement of sustained crop productivity, especially in intensive cropping systems, depends primarily on the maintenance of soil fertility. Farmers have developed four approaches to the problem: purchasing commercial fertilizers, collecting nutrient materials from outside the farm, recycling nutrients from a single cropping system, and recycling nutrients among various cropping systems on the same farm. Three of these approaches involve the importation of nutrients into the field from external sources.

Purchased fertilizers

On a commercial farm, the modern approach to maintaining soil fertility is straightforward enough: buy fertilizer. If the farmer has both access to commercial fertilizers and a ready cash market for his crops, purchasing plant nutrients is a normal and profitable investment in his highly productive farming system.

In the market-dominated agriculture of developed countries, the purchase of commercial fertilizers derived from industrial chemicals or recycled urban wastes is a standard procedure. The farmer can precisely tailor his fertilizer purchases and applications to the needs of his soil and crop. This method of fertilizing the soil requires a minimum of labor to eliminate nutrient deficiencies as limiting factors in intensive crop production. If the small farmer is involved in the cash economy, and commercial fertilizers can be made available to him, he can use the same method; but especially for the small farmer, reliance on commercial fertilizers should be only one of a variety of approaches to the problem of soil fertility.

Nutrients collected from outside the farm

A second approach to improving soil fertility uses nutrients found outside the farm. In the hills of Nepal, where farmers graze their animals on communal lands, they carefully collect the manure from the pastures for composting with manure from the enclosures in which the animals are kept at night. This organic material is supplemented with leaves and other plant materials collected from nearby forests. Used to fertilize intensively cropped fields, the compost substitutes for commercial fertilizers, which are not generally available.

The productivity of farms fertilized in this manner depends on the number of animals producing manure on each farm, access to communal grazing lands to feed the animals, and the proximity of forests to provide vegetable matter. The system is efficient in making use of nutrients from sources that would otherwise be wasted, but it is limited by its requirements for labor and by its requirement for a fairly fixed ratio of cropland, forestland, and grazing land. Moreover, the system can easily become overtaxed by population pressure, which reduces the available open land while increasing the demand for food production.

Recycled farm materials

A third fertilizing practice recycles nutrients among the various enterprises on the single farm. Waste plant materials are composted with wastes from the household, especially ashes from the cooking fire. Fodder from fencerows, crop residues, and other sources is fed to livestock, and the resulting manure is added to the compost. The compost is allocated to the crops according to their value: vegetables, for example, might be first in line to receive fertilizer.

Cycling nutrients among farm enterprises in this manner, like importing fertilizer from off-farm sources, is labor-

Composting animal and farmyard wastes is one way small farmers recycle nutrients efficiently.

intensive. Moreover, if the process is not carried out well under favorable conditions, the compost may lose much of its nutrient value from leaching as it is exposed to rain. The efficiency of the recycling system is improved if animals are tethered in the fields during the dry season, dropping their wastes directly on the soil with no loss of nutrients and little human labor. In northern India, the sheep flocks of semi-nomadic herdsmen are invited into fields by local farmers who want the manure to fertilize their next crop.

Adoption by farmers of nutrient recycling systems seems to be related directly to the cost of commercial fertilizer and inversely to the opportuniy cost of labor. Thus, when labor, on- or off-farm, has a relatively low cash value, or fertilizer is scarce and expensive, nutrient recycling becomes an attractive alternative, as illustrated in Figure 12.

In addition, the farmer's basic decision to invest either his labor or his cash in any system of soil nourishment depends on both the prospective market value of his crop and its potential response to additional nutrients. If a kilogram of fertilizer is equal in value to 5 kilograms of the crop, the farmer may look for a crop response of at least 10 kilos of yield per kilo of fertilizer before he decides that the investment is worthwhile.

Nutrients recycled within each crop

The fourth approach to maintaining soil fertility exploits the crop's ability to reuse its own stored nutrients. The paradigm for this kind of self-nourishment is the forest ecosystem, in which nutrients are withdrawn from the soil over long periods of time and stored in the biomass of living plants. Forest plants are equipped with penetrating root systems, which are very efficient at extracting nutrients from deep in the soil profile. The preponderance of long-lived, woody species enables the ecosystem as a whole to survive the dry seasons and to take advantage of the first rains of the wet season, that flush accumulated nutrients from the soil surface down to the root level. In this way the nutrients accumulated in fallen leaves and branches, entire plants and trees, and the manure of forest animals are largely recycled through the system.

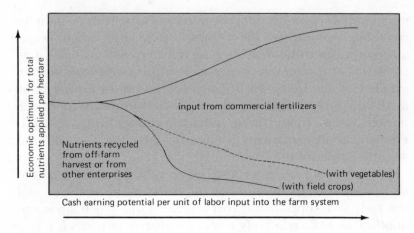

Figure 12.
Cash returns to labor in relation to method of meeting crop nutrient requirements.

The same closed-system recycling takes place in large fieldcrop monocultures, but only to a very limited extent. Crops that mature quickly are far less efficient absorbers of nutrients from deep in the soil profile than slow-growing forest plants in the first place; and much of their accumulated nutrients is lost to the system in the harvest.

In traditional slash-and-burn agriculture, nutrients stored in the forest ecosystem are rapidly released and recycled for agricultural use. As the forest is burned to clear the land for planting, nutrients are immediately released in highly soluble form, available to support crop growth. On the conventional small farm, however, nutrients are applied when compost is moved to the field during the farmer's slack periods, when he places a lower value on his labor. It is important to understand this nutrient cycling process in order to select appropriate measures to achieve more efficient nutrient use.

The basic process is illustrated in Figure 13. By a physical and chemical weathering process, nutrients are released as the parent material breaks down over long periods of time and soil is formed. As they dissolve, the nutrients become constituents (*1*) of the soil water (*D*). The concentration of nutrients in the soil water is very low, in equilibrium with the nutrients attached to clay particles in the soil itself and thus relatively

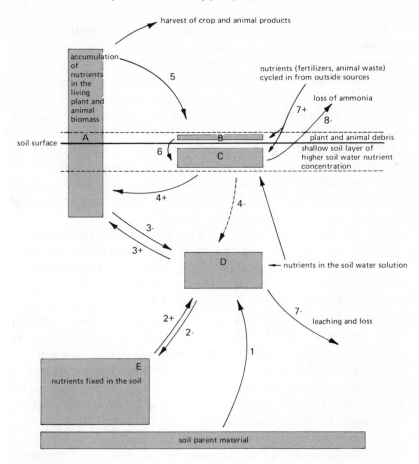

Figure 13.
Nutrient cycling in cropping systems.

unavailable to the plants (*2, 2-*). The plants absorb nutrients from the more readily available supply in solution in the soil water (*3+*) and accumulate these as plant biomass (*A*) in the growth process. When the plant, or part of the plant, dies, its biomass is deposited on the soil surface (*5*), where nutrients accumulate. This plant refuse, along with a portion of the living plant material, is broken down by biological processes involving soil microflora into organic acids and soluble nutrients (*6*), which are dissolved into the soil water in the upper few centimeters of soil (*C*).

Nutrients added to the system—through applications of fertilizer, recycling, flooding, or other mechanisms—are absorbed in the material on the soil surface (*B*) or dissolved in the upper soil water (*C*). The nutrients in the upper soil water (*C*) then travel (*4-*) to the lower soil water supply (*D*) where they are absorbed by growing plants (*4*) until the nutrient concentration in the soil water comes into equilibrium with that in the soil itself (*E*). Some nutrients are lost through the leaching process (*7-*); nitrogen is lost through ammonification and the escape of ammonia gas into the air (*8-*), through denitrification and loss of nitrogen compounds, or through burning and direct loss to the atmosphere.

The relative concentrations of nutrients in the plant biomass (*A*), the soil surface (*B*), the upper soil water (*C*), the lower soil water (*D*), and the soil itself (*E*), determine the productivity of the system. They are also the key to nutrient management under different resource conditions. For the rapid growth and high yield of an improved variety of an annual crop, nutirents must be concentrated in the soil water (*C, D*) immediately adjacent to the plant roots. The maintenance of this nutrient concentration is the specific aim of all fertilization programs.

In high-fertility soils, the concentration of nutrients in the soil water (*D*) is maintained at a high level by the natural movement of nutrients from less available forms. In such soils, crops may show little response to the addition of fertilizers. When soils are naturally deficient in one or more of the major elements necessary for plant growth, on the other hand, it is usually impossible to add enough fertilizer to raise the general nutrient level of the soil sufficiently to provide good support for crops. The objective of an enrichment program is to apply the needed nutrients as close to the plant roots as possible at precisely the time when the crop most needs them. Even then, the crop's absorption of the applied nutrients will seldom exceed 40 percent and will often be considerably lower.

Each of the traditional techniques of soil enrichment described above can be used as an efficient way of providing nutrients to crop plants under conditions of low soil fertility.

In stable intensive-cropping systems, deep-rooted perennial plants capture nutrients from deep in the soil profile and incorporate them in their substantial biomass. These nutrients are recycled to the upper soil layer as the crop drops leaves or is cut and mulched. In Indonesia, the deep-rooted, erect-branched, leguminous tree *Gliricidia maxima* is commonly planted in rows or on the borders of rice paddies. It provides green manure, animal feed, and firewood, as well as flowers for human consumption. Such a system is most efficient when organic material cannot readily be imported from outside the farm, and economic tree species can be used as a source of soil nutrients. The recent interest in *Leuceana glauca* is due to the exceptionally high nitrogen content of its leaves. Despite this virtue, however, the tree is characterized by spreading branches which compete with crop plants for sunlight. Moreover, the present practice of strip-cropping the trees and cutting them when they are small may consume too much labor and result in weed problems in the *Leuceana* growing areas. A better alternative would seem to be the *Glyricidia*, pruned in the dry season and its larger limbs burned.

A second method of recycling nutrients efficiently involves the careful use of plant residues by mulching them, rather than burning them off or plowing them under. Mulching permits the gradual breakdown of organic materials and the concentration of nutrients in the upper soil layers over a period of several weeks. This process makes the nutrients available just when the next crop is reaching its peak nutrient demand. This method is coming under intense scientific scrutiny in West Africa.

The third alternative, also involving the use of a stubble mulch, is based on highly mixed intercropping. Especially when plant refuse has been left on the surface during the dry season, the availability of nutrients in the upper soil layers increases markedly with the onset of the monsoon rains, then decreases gradually through the wet season as the nutrients are absorbed or leached, reaching its lowest level in the dry season. This schedule perfectly suits a mixture of crops of different maturities, ranging from 2 to 10 months, planted at relatively low densities. Such a mixture will have a relatively high nutrient demand early in the wet season because of the needs of fast-growing crops, such as maize. The nutrient demand will decrease through the season, however, as the longer-duration annual crops in the mixture reach maturity. Thus, the crop's nutrient demand is synchronized with the nutrient availability in the soil.

It should not be inferred, however, that any of these practices will match applied fertilizers in producing high yields. Such traditional methods are only relatively efficient ways of using existing sources of plant nutrients to produce modest yields when commercial fertilizers are unavailable or too expensive.

It is important that soil scientists in the developing countries give full consideration to these and other traditional farmer practices that promise to increase the efficiency of soil enrichment when commercial fertilizers are not a realistic alternative. Expensive fertilizer inputs, even when they are available, can be combined with traditional nutrient cycling methods to increase the economic efficiency of the small farm system. No new major research programs are needed; only a broadening of professional interest in existing programs and in traditional farmer technology.

Efficient Use of Farm Resources

On small or remote farms, many agricultural methods have evolved in response to specific resource limitations. Methods for increasing the efficiency of purchased inputs are of particular importance to the small farmer, because the cash with which to purchase these inputs is often his most limited resource. Therefore, the farmer values any method that will make his purchased inputs go further, or that will make them unnecessary.

Farmer priorities for resource use

Agricultural development professionals generally agree that the farmer must attain some specific level of general well-being before he will be willing to invest his scarce discretionary income in an effort to increase his farm production. In Figure 1, that level of well-being was defined as being reached when the farmer's productivity raised his family's standard of living above the hunger threshold. Until that level of family welfare is attained, a subsistence farmer's priorities for his meager cash income may well include food, clothing, and a few minor consumer goods before increased investment in fertilizers, herbicides, pesticides, machinery, or other production-related goods.

When a farmer does begin to buy production inputs in the retail market, he will tend to spend his scarce cash first for those inputs that most severely limit his production and which he cannot secure from other, noncommercial sources. Fertilizer is usually high on his shopping list. Herbicides, which can be perfectly well substituted for by hand labor, with or without animal or mechanical power, are usually low-priority purchases.

Fertilizer

On intensively cropped small farms where labor is plentiful, as in Java, fertilizer is usually the major purchased input, carefully dibbled into the soil beside each plant at planting time and again later in the growing season. Such painstaking and timely fertilizing is labor-intensive but efficient in the use of a scarce resource—fertilizer—at the expense of one that is relatively cheap—labor. This practice would be impossible on large commercial farms, where labor is more expensive.

Most agricultural development programs feature increased investment in commercial fertilizers to improve the productivity of target crops. It is not uncommon in such programs, however, to find farmers diverting their scarce fertilizer from the target crop to one that is not included in the development program. In most of these cases, the farmer is making his own decision about his priorities. Often, he is applying his cash resources to the crop that will yield the highest return on his investment.

Farm power

The kind and amount of animal or mechanical power available to the farmer are critical factors in the efficiency of his agriculture. The balance between mechanical or animal power and human labor that is needed to produce the highest efficiency is illustrated in Figure 14. For intensive cropping, a relatively small power unit is preferable because it can economically service small areas, enabling the farmer to schedule his power use according to the precise requirements of his different crops. Often, a single draft animal is the most appropriate power source. In other situations, however, especially those in which labor is scarcer than power and a single crop or a two-crop sequence allows flexibility in scheduling, a large tractor of 75 horsepower or more may be the most efficient power source.

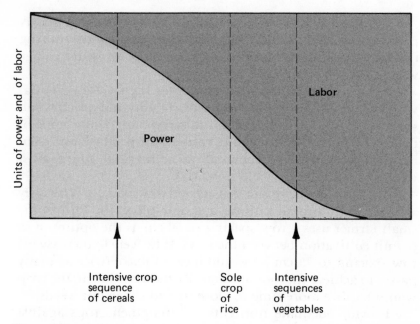

Figure 14.
Power-labor ratios needed for various crop enterprises.

Where labor and feed are available, draft animals conserve cash resources, requiring small initial investments and minor maintenance expenditures. Only on farms with adequate capital and cash flow and a high power demand for tillage is mechanical power more efficient than animal power.

Crop diversity and management

The farmer's selection of crops and management methods also affects the efficiency with which farm resources are employed. Some crops respond only moderately to high levels of management. Others, such as rice, respond more generously to management, but the style and magnitude of the response varies widely among the different varieties of the crop; the farmer must select the particular variety that is most appropriate to his level of management as well as to his agroclimatic conditions. In addition, the farmer can take a number of specific steps to increase the efficient use of his resources:

• Devote sufficient power to tillage before planting. A poorly prepared seedbed usually costs more for weed control during the growing season than it would have cost for better tillage when the land was clear.

• Plant a crop variety that combines high yield potential with an aggressive response to weeds. Traditional varieties and some improved varieties exhibit this response, while some of the highly touted high-yielding varieties are prima donnas that require tender loving care, including careful protection against weeds, to realize their yield potential.

• Maintain proper plant density and row spacing. With only a single draft animal pulling a simple plow, the Philippine small farmer uses a row spacing of 60 cm as the optimum to permit cultivation between the rows. If he were to increase his row spacing to 75 cm he would have to make twice as many passes to achieve thorough tillage. By the same token, the crop would require more time to close in and shade out weeds.

• Provide adequate nutrients to strengthen crops against weeds. Properly nourished, a crop like maize will be far more competitive with weeds, requiring less labor and cash investment for weed control. Other crops, such as mung bean, show less response than weeds do to nitrogen. Such crops therefore require somewhat less nitrogen for optimum weed control.

In intensive farming systems, the critical factor in the efficient use of scarce resources is diversity. As any resource becomes scarce enough to limit production, diversity becomes increasingly important to its optimum utilization. In the course of a year, each farm enterprise makes uneven demands on various farm resources—power, labor, money, nutrients, or water. If labor, for example, is the first limiting factor, the shortage will first occur when the crop makes its peak labor demand—at planting, weeding, or harvest. The farmer can increase the efficiency of his labor and power use in a single-crop, intensive sequence by planting varieties of differing growth periods, by using a variety of planting methods, and by planting at different times.

Requirements
for Farm Mechanization

A farm's power requirements increase dramatically as cropping is intensified. This increased appetite for power is not by itself an argument for the mechanization of small farms. On farms of less than one-half hectare, a high level of cropping intensity can be sustained with only the use of human labor. Animal power is sufficient on farms up to 2 hectares in size, especially when the soil is tilled wet.

Primary mechanization

As cropping intensifies, however, so do the time constraints on the application of labor or power, especially for primary tillage. To grow two or more crops during the months when water is available, the farmer must not only increase the number of his tillage operations but he must also carry them out on a much more precise schedule than he would if he were growing only a single crop. He no longer has the luxury of one or two months to prepare his fields for planting. He may have no more than two or three weeks from the harvest of his first crop to the planting of his second. His timing problems are compounded by the heavy soils typical of most tropical farms and the frequent rains that fall in the short period between crops.

These characteristic conditions on intensively cropped farms make certain special demands of the farmer's power source. In general, the source must be as small as possible while still producing enough power to do its assigned work. In areas where farms are of less than 5 hectares, a single power source can economically service no more than a dozen farms. A larger unit, servicing more farms, will not be able to meet the critical timing requirements of intensive cropping sequences. Larger units can be safely used only at the start of the rainy season, when land preparation, especially the initial deep plowing, can be spread over a month or two. When his operations must be more closely scheduled, however, the farmer cannot depend on the timely availability of power that is too widely shared with his neighbors.

A smaller power unit can be more easily purchased by a single farmer or a small group, and it can be used efficiently to service a small area on an intensive-cropping schedule. A tractor of 5 to 7 horsepower, for example, can probably be justified for primary tillage of only 10 to 15 hectares per year. A farmer who grows two to three crops per year on 3 hectares might buy such a small machine for his own primary tillage, seedbed preparation, and other uses. If he were to buy a bigger tractor, however, he would have to contract it out to other farmers to make it pay for itself.

Small tractors of less than 10 horsepower are most useful in lowland paddy conditions, where less size and power are sufficient to do the plowing. Under upland conditions, however, such small machines are useful only for small-scale garden farming in which labor is available to clear the field of crop residues before plowing. For field crops in upland conditions, a tractor of from 12 to 15 horsepower seems to be about the smallest that will do the job.

The kind of tillage to be done also affects the size of the power source. For rotary tillage with disc plows, large tractors of 50 to 75 horsepower are now commonly used. This combination of tractor and plow has many advantages over the older, moldboard plow in conditions of heavy, wet soil, imperfectly cleared fields, or heavy crop or weed stubble. In addition, rotary tillage with large tractors works well under many lowland conditions and on well-cleared fields. In combination with smaller tractors, the fixed moldboard plow is more effective than a disc plow on fields with crop residues and stubble. Improved rotary equipment for tractors of 12 to 15 horsepower is being developed rapidly, however, and is already being used extensively in Japan and Taiwan.

In short, then, power for primary tillage becomes increasingly necessary as cropping intensity increases. On farms of less than 5 hectares, tractors of 50 horsepower or more are useful only during the initial transition to intensive cropping. As cropping intensifies, smaller power units that can service smaller areas on a tight schedule are more efficient. Tractors of 12 to 15 horsepower equipped for rotary tillage are most effective under upland conditions and for upland-lowland rotations. The effectiveness of rotary-tillage equipment, especially in combination with a small power source, is increased by clearing the field of crop residues before plowing.

Secondary mechanization

Secondary mechanization includes all farm uses of mechanical power other than primary tillage. The most common and profitable application of secondary mechanization is in water pumping, which has an immediate beneficial effect on crop productivity. Pumps cost relatively little to buy, and they are also cheap and simple to maintain. They represent an efficient use of the farmer's scarce cash resource.

In the process of farm development, illustrated in Figure 1, secondary power in the form of a pump often precedes primary power in the sequence of capital investments that a farmer makes. Pumps are available in every size from the very small to the very large, so the farmer can choose precisely the scale of technology appropriate to his circumstances.

After water pumps, machines to thresh cereal crops are probably the next items on farmers' lists of priorities for mechanization. Threshing is especially labor-intensive and arduous, and the process is a common bottleneck in an intensive cropping system. The available machinery for threshing is often too large for the small farmer to own individually or with a small enough group of neighbors to permit practical sharing. Small, efficient threshing units are now becoming available, however, and many of them can share a pump motor or be belt-driven by a small tractor engine. As increased cropping intensity forces increases in labor productivity, mechanized threshing will become more common. Where land is limited and crop production is static, however, the mechanization of threshing will simply displace labor.

Where grain is grown in vast monocultures as a commercial crop, mechanized threshing on a large scale is practical, even essential. The threshing operation is usually done under contract by businessmen, often grain dealers who thresh, ship, and store the farmer's crop. Contract threshing is most practical with crops that can be stored between harvest and threshing, such as maize and—in some cases—rice. Cassava can be harvested when needed—stored in the ground, so to speak—so it falls into the same category when it is to be chipped and dried. A crop like sorghum, on the other hand, which can only be stored after harvest under extremely dry conditions, often will not keep long enough to be handled in a large-scale contract threshing operation. Crops that are largely consumed by the farm family, or that are sold in small amounts in local markets, or that are not widely grown, cannot be threshed efficiently in large-scale contract operations.

The most difficult function to mechanize on a small tropical farm is cultivation, either for weed control or to break up the soil surface so water can infiltrate after the crop has emerged. The prerequisite for mechanized cultivation is precise, mechanized planting that results in perfectly even row spacings. Such precise planting requires in turn that the field first be smooth and absolutely free of obstacles. It also requires large machinery, including expensive seeding equipment. The seed must be graded and well cleaned. Row spacing, ridge shape, and furrow shape must all be matched to the equipment with a precision that is entirely unknown in cultivation with animal or human power. Cultivation equipment is also quite particular about soil moisture, performing best under the kinds of ideal conditions that are rare in the heavy soils and frequent rains of the tropics. Finally, mechanized cultivators tend to be too large to turn around in small fields bordered by hedgerows or bunds.

For all these reasons, there is little mechanized cultivation of crops on small farms in the tropics. In the portions of Thailand

and the Philippines, primary mechanization is used in a two-tiered agriculture in which farmers hire independent contractors for mechanized primary tillage in a mixed animal-and-crop farming system. Draft animals are used for subsequent crop cultivation, giving the farmer the advantages of small-scale operations to meet the precise scheduling requirements of his cropping system. Chemical herbicides are also used to control weeds, reducing the need for cultivation during the growing season.

Transportation

A final application of farm mechanization is for on-farm transportation. A farmer's fields are often widely separated, requiring considerable hauling of materials, produce, and people. As labor becomes too valuable to be used for manual hauling, mechanical power takes its place. Whatever power source the farmer adopts, therefore, should be readily adaptable for hauling.

Power and farm resource use

The need to spread the demand for the farmer's limited labor and power resources increases as cropping intensifies. To make the most efficient use of his power source, the farmer is well advised to invest in a power unit no larger than is required by the physical demands of the job, and then to work his machinery as continuously as possible in a planned sequence of crops. Both the type and timing of the crops can be adjusted to make the best use of the power source. If heavy rains promise to make tillage difficult in a particular month, for example, the farmer can time his planting so that neither land preparation nor harvest will occur at that time. Similarly, the crops should be scheduled so that the power source is available when it is required. The availability of irrigation water, as well as other factors, may force the farmer to alter his schedule, but whenever possible the rest of the farming system should be integrated with the power source.

As mechanization progresses, the farmer's need for commercial and technical skills increases sharply, as illustrated in Figure 1. At the same time, there is a rapid increase in the need for supportive services from the community. These services, the farmer's skills, and the mechanization process must proceed at the same pace.

The process of mechanization requires the farmer to be involved in a market economy and to generate sufficient cash flow to afford the capital investment and operating costs that are entailed. The farmer must already enjoy a standard of living high enough to allow him to divert some of his income toward the costs of mechanization. Until he has reached a sufficiently high level of well-being, he is likely to have other uses for his income that will preclude investments in mechanization.

Beyond this economic threshold, the farmer will mechanize if he can afford the costs and if the investment promises to pay. He will seldom choose the luxury of an inefficient investment in power machinery merely to eliminate drudgery. Rather, he will tend to make a rational economic decision between buying machines and hiring labor.

Mechanization is often accused of displacing farm labor, idling rural workers and forcing them to migrate into overcrowded cities where unemployment is already high. To avoid these undesirable side effects, the development planner must judge mechanization case by case. Mechanization can be justified if it will increase farm productivity by improving tillage to control weeds and foster crop growth, or if it will make crop intensification possible where labor is scarce and expensive. Under these conditions, mechanization can supplement and amplify labor while increasing productivity. On the other hand, if productivity does not increase, increased labor efficiency through mechanization will simply reduce labor requirements, causing unemployment where labor is abundant. In such a case, mechanization is a substitution of capital and mechanized energy for human labor. It is essential, therefore, that planning and development specialists be intimately familiar with the capital and labor circumstances of the farmers with whom they are working so they can make informed predictions of the effects that different degrees of mechanization will have. The special insights of economists and production specialists are essential to this analysis.

Stability in Farming Systems

Farmers with limited production resources want to be as sure as possible that their investment of those resources will pay off in substantial production increases. When agricultural development specialists urge a farmer to adopt a particular crop on the basis of its performance in past trials in similar environments, the farmer must decide whether the predicted performance of the crop on his land is worth the risk of a deviation from its past performance. He wants to know how stable the crop is in living up to its promise. The stability, or predictability, of a cropping system and its various components is of the utmost importance to the farmer faced with a decision whether to adopt a new technology, and this factor must therefore be considered by those who design new farming systems.

Stability is inseparable from risk, a familiar economic concept that figures largely in farmers' decisions. The total risk in any proposed innovation is an aggregate of several factors that can be analyzed and dealt with individually. Each is an element of uncertainty, instability, or unpredictability in some aspect of the production system. As farming systems are designed, it is important that stability be built into each element. Factors that will increase the farmer's risk—and thus make the entire system less acceptable to him—must be identified in the planning stage and corrected as much as possible to produce a stable system.

There are various sources of stability in farming systems.

Biological stability

The biological stability of a crop—plant or animal—is the measure of its ability to deliver a predictable yield under given environmental and management conditions. A crop with a high degree of biological stability will maintain its expected productivity despite fluctuations in weather conditions and disease and pest incidences.

In the humid tropics, where rainfall is plentiful but erratic, most animal enterprises and tree crops have greater biological stability than annual crops because trees and animals are less affected by short-term weather fluctuations and less susceptible to pests and diseases. In the same environment, a crop like onion has low biological stability because it is very sensitive to water supply. Lowland paddy rice is much more biologically stable than upland rice.

The biological stability of a crop can be increased by breeding and selection to improve such inherent qualities as drought tolerance and resistance to pests and diseases. The new tropical varieties of wheat, rice, and maize have inbred resistance to a broad spectrum of insects and diseases; many also have a genetic tolerance for drought and adverse soil conditions. The deepwater dwarf varieties of rice from Thailand, which elongate their stems rapidly in response to increased water depth, are an excellent example of genetic adaptations that increase the biological stability of a crop for use under uncertain environmental conditions.

The biological stability of a crop can also be increased by proper management. The choice of the optimum time for planting, the proper use of fertilizers, adequate field drainage, and other management practices can markedly improve the predictability of crop performance.

Biological stability can also be affected by the diversity of crops. In areas where farmers plant patchworks of different crops in adjacent fields, the biological diversity of the total system usually deters devastating outbreaks of pests or diseases. Severe infestations seldom occur in highly diversified, mixed farming areas.

Such diversification is not always practical in areas devoted to seasonal monocultures of rice, maize, sugarcane, and other staple, commercial crops. Diversity can be introduced, however, by planting varietal mixtures of the single crops, including varieties with different genetic resistances to pests and diseases. Recent research has also demonstrated the efficacy of mixed cropping and intercropping in reducing pest and disease damage, thus increasing biological stability.

In selecting the crops for his total enterprise, the farmer often considers the contribution of diversity to biological stability and aims for the maximum overall stability in his system. If he has ready markets for more than one high-value crop, for example, the farmer may forego the increased theoretical efficiency of a single crop in favor of a mixture of crops with complementary sources of biological stability. A farmer might retain coconut trees that are not as profitable as an alternative crop but that are extremely stable, affording him a hedge against a failure of his more profitable but less stable crop. He may also grow cassava, a very stable annual crop; upland rice, which is less stable; and some vegetables, which are highly unstable but potentially highly profitable. One or more animal enterprises add extra stability to the total system. The achievement of a balance among crops with different degrees of biological stability is an important motivation for farmers as they make decisions about intensive mixed farming systems.

The biological stability of a crop is reflected in the costs the farmer will have to bear to grow it. Highly unstable crops, which are often highly profitable in the end, also often cost more to produce than more stable crops because expensive inputs must be used to compensate for their instability. A crop that is biologically unstable because it is extremely susceptible to insect depredations, for example, must be protected with pesticides or other costly measures.

Management stability

A second source of crop stability—one that is commonly overlooked or misunderstood by development planners—is the ability and readiness of the farmer to carry through the appropriate management program that will ensure the success of the crop. Many otherwise promising production schemes have failed because of management-induced instability— although the failures are almost always ascribed to lack of management.

A clear example of management-induced instability is the combination of mechanical cultivation and hand weeding of upland rice in parts of the Philippines. The method works well under ideal conditions of moderate to low rainfall. When the rainfall is unusually heavy during the first month after planting, however, the farmers cannot get their mechanical equipment into the fields on time and, when they do, the wet soil reduces the effectiveness of the cultivation.

Moreover, when this upland management method was modified for direct-seeded lowland rice, a weed control chemical was added in an effort to improve the stability of the crop. The chemical's effectiveness was short-lived, however, while the paddies remained unflooded for up to two months. As a result, the resurgent weeds had a chance to outgrow the dwarf rice. Production on many farms has suffered greatly from this failure of management to improve the stability of the crop. Alternative management methods to increase stability include:

- planting direct-seeded rice only in paddies that can be flooded within 30 days
- using rice varieties that are more competitive with weeds
- using an herbicide with a longer active life
- direct-seeding only in paddies where weed control has previously been good

Thus, as the Philippine example illustrates, the farmer has available a variety of management options that will improve crop stability. Many of these options are pest management methods, which are critical elements in management stability and in the overall success of any crop enterprise. Weed management is perhaps the most important single element needed to improve crop stability in the humid tropics, and there is an urgent need for new crop varieties that can compete successfully with weeds, as well as for other innovative and effective weed management strategies.

In deciding whether to adopt a new management technology, the farmer gives great weight to the stability it will lend his crop. An unstable management technique or package will result in a crop that is difficult to grow, involving techniques that are impractically complex or that leave the crop vulnerable to environmental factors. The high probability of failure with such a management system is likely to discourage the farmer from adopting it. To the farmer, failure may mean not only loss of crop production, but loss of respect as well. It is important for the development planner therefore to recognize the management instability in new technologies and the devise alternative management strategies to eliminate or compensate for it.

Production stability

The overall production stability of a farm is the result of the biological stability and the management stability of each of its component enterprises. The importance of production stability to the farmer depends largely on his economic circumstances. A farmer with extremely limited resources, facing anew with each crop the absolute need for at least a minimum production, cannot afford to endure any more instability than the irreducible minimum. He cannot afford to take a chance on an unstable crop or management technology, even if it promises to repay increased risk with increased production. He will demand the utmost stability on the subsistence portion of his farm; on whatever portion is left for commercial enterprises, he will be more willing to accept some instability for the chance to make money.

Economic stability

In commercial farming, economic stability is a combined measurement of production stability and price stability. The farmer's inability to predict the market prices for his crops, especially when he must also contend with biological and management instabilities, adds to his reluctance to adopt new technologies. Other economic factors—the advantage of off-farm income, for example, or the burden of debt—can also affect the economic stability of the farm as a whole.

In general, the farmer makes trade-offs between productivity and stability. Many traditional farming systems have evolved a high degree of stability at the cost of only modest productivity losses. As the farmer intensifies his cropping, pushing his resources toward their theoretical limits, he sacrifices some measure of stability. "Nothing ventured, nothing gained," as the old adage puts it.

Nevertheless, carefully designed technologies can combine high productivity with considerable stability. The development of alternative technologies affords the farmer a choice of approaches to his particular needs and circumstances. The rice farmer mentioned in chapter 8, for example, was willing to try the unstable direct-seeding technology with its potential for high yield largely because he had several alternative management technologies to rely on if it failed.

As new technologies are developed for intensified farming, each must be carefully assessed for stability. The biological stability of a new crop variety can be reliably estimated from environmental data combined with performance results of the crop in trials under similar conditions. Management stability, on the other hand, can only be assessed under actual farm conditions. Before a new management technology is recommended to farmers, therefore, it is crucial that it be tested under farmer management. Failures should be analyzed carefully to identify the exact source of the instability.

The importance of farmer participation in the testing of new technologies can hardly be exaggerated. There is a strong temptation to run on-farm tests under the management of researchers to ensure proper controls, but such trials do not reflect the true management characteristics of the technology. In the final analysis, it is the farmer who will manage the crop.

Appendixes

Appendix A:
Sources of
Farming Systems Information

Development leaders seeking to begin or strengthen work in farming systems improvement can profit from the experiences of programs already under way. This brief listing of such programs has been divided according to the kinds of information they can best provide. This division is a partial reflection of their research orientation.

A qualitative understanding of specific farming systems may be obtained from current anthropological and geographical literature. A few programs involved in such studies include:

Bernice P. Bishop Museum
1355 Kalihi St.
P.O. Box 6037
Honolulu, Hawaii 96818 USA

University of East Anglia
School of Development Studies
Norick NOR 88c
Norfolk, England

University of Hawaii
Department of Geography
Honolulu, Hawaii 96822 USA

Quantification of the main component effects of specific farming systems. These data are available from much of the current economic literature concerned with tropical agriculture. Cost and returns of individual commercial crop or animal enterprises constitute the major portion of these types of data and are available for most commercial enterprises, usually with only scant reference to the farming systems or environmental contexts.

Quantification of entire systems by scientists who "observe" only.

Michigan State University
Department of Agricultural Economics
East Lansing, Michigan 48823 USA

Iowa State University
Department of Agricultural Economics
Ames, Iowa 50011 USA

International Crops Research Institute for the Semi-Arid Tropics
1-11-256, Begumpet
Hyderabad, 500016 (A.P.), India

International Institute for Tropical Agriculture
PMB 8320
Ibadan, Nigeria

Asian Vegetable Research and Development Center
P.O. Box 42
Shanhua, Tainan 741
Taiwan

Farmer-participant on-farm research of entire systems with standard treatments (of crops) included in the systems from the outset. The entire system is monitored for measuring interactions with test components.

Centro Agronómico Tropical de Investigación y Enseñanza
Turrialba, Costa Rica

Cropping Systems Program
International Rice Research Institute
P.O. Box 933
Manila, Philippines

The Multiple Cropping Program
Central Research Institute for Agriculture
Bogor, Indonesia

Division of Soil Fertility
Department of Agriculture
Bangken, Bangkok
Thailand

Multiple Cropping Project
Bangladesh Rice Research Institute
P.O. Box 911
Dacca, Bangladesh

Quantification of the interactions within systems through farmer participant research and measuring their changes across environmental gradients, permitting prediction of systems behavior and design of systems to fit target environments.

The International Rice Research Institute
P.O. Box 933
Manila, Philippines

Centro Agronómico Tropical de Investigación y Enseñanza
Turrialba, Costa Rica

Farming systems extension methods

University of the Philippines
College of Agriculture
Los Baños, Philippines

Appendix B:
Farming Systems Terminologies

The attention focused on farming systems technology over the past decade has created a rapidly expanding vocabulary. Disagreement has concurrently arisen over older terminologies. The more commonly accepted definitions are presented here and major areas of disagreement are indicated. Where more than one definition appears, the usage in this book is listed first.

Terms relating directly to systems as a whole

System. An assemblage of objects and activities united by some form of regular interaction or interdependence.

Farming system. The manner in which a particular set of farm resources is assembled within its environment, by means of technology, for the production of primary agricultural products. This definition thus excludes processing beyond that normally performed on the farm for the particular crop or animal product. It includes farm resources used in marketing the product. Another definition is "a collection of distinct functional units, such as crop, livestock and marketing activities which interact because of the joint use of inputs they receive from the environment." This more general use of the term thus transcends individual farm boundaries, referring to similarly organized farm units.

Cropping system. The cropping patterns used on a farm and their interaction with farm resources, other farm enterprises, and the technology which determines their makeup.

Farm enterprise. An individual crop or animal production function within a farming system which is the smallest unit for which resource

136

use and cost-return analysis is normally carried out, i.e., the raising of a particular kind of animal or a single planting on one date of a particular crop. An enterprise is thus a subsystem of crop or animal systems and of the farming system as a whole.

Shifting cultivation. Several crop years are followed by several fallow years with the land not under management during the fallow. The shifting cultivation may involve shifts around a permanent homestead or village site, or the entire living area may shift location as the fields for cultivation are moved.

Slash and burn. A specific type of shifting cultivation in high rainfall areas where bush or tree growth occurs during the fallow period. The fallow growth is cleared by cutting and burning.

Dry farming. The cultivation of cereals in rotation with one or two years of fallow in arid and semi-arid zones.

Rainfed farming. The growing of crops or animals under conditions of natural rainfall. Water may be stored in the crop field by bunding, as with lowland rainfed rice, but no water is available from permanent water storage areas.

Mixed farming. Farms with integrated crop and livestock activities.

Silviculture. The growing of trees for lumber or other wood products.

Agri-silviculture. The growing of trees for timber but with cultivated crops grown beneath.

Terms relating to the type of crop

Crop. All plants on a farm which are planted and managed for economic purposes, producing a physical product for farm use or sale.

Arable crops. Crops requiring cultivation.

Short-term crops. Crops occupying land for three months or less.

Medium-term crops. Crops occupying land for three to six months.

Long-term crops. Crops occupying land for 6 to 18 months.

Perennial crops. Crops occupying land for more than 30 months (not including legumes and grasses in permanent pasture).

Perennial field crops. Crops which require cultivation and which occupy the field for 3 to 12 years (sisal, sugarcane).

Shrub crops. Trees which are made to develop a shrub-like appearance (coffee, tea).

Tree crops. Trees yielding fruits and not primarily grown for timber.

**Terms relating to the spatial
and temporal arrangement of crops**

Multiple cropping. Growing more than one crop on the same land in one year. Within this concept there are many possible patterns of crop arrangement in space and time.

Cropping pattern. The yearly sequence and spatial arrangement of crops or of crops and fallow on a given area.

Cropping index. Number of crops per year on a given field multiplied by 100. (Sometimes used as R-value, the percentage of crop land actually cropped in a year.)

Land equivalent ratio (LER). The area needed under sole cropping to give as much produce as 1 hectare of intercropping or mixed cropping at the same management level, expressed as a ratio. LER is the sum of the ratios or fractions of the yield of the intercrops relative to their sole-crop yields.

Area-time equivalency ratio. The ratio of number of hectare-days required in monoculture to the number of hectare-days used in the intercrop to produce identical quantities of each of the components.

Maximum cropping. The attainment of the highest possible production per unit area per time without regard to cost or net return.

Sequential cropping. One crop is planted after harvest of the first.

(Sometimes called relay planting in West Africa.)

Monoculture planting. Growing a single crop on the land at one time. Another definition is "the repetitive growing of the same crop on the same land."

Sole crop. One crop variety grown alone in pure stands at normal density.

Ratoon cropping. The cultivation of regrowth from stubble following a harvest not necessarily for grain.

Double cropping. Growing two crops in the same year in sequence, seeding or transplanting one after the harvest of the other (same concept for triple cropping.)

Strip cropping. Growing two or more crops in different strips across the field wide enough for independent cultivation. The strips are wide enough to give greater intra-crop than inter-crop association.

Intercropping. Two or more crops grown simultaneously in the same, alternate, or paired rows in the same area.

Interplanting. All types of seeding or planting a crop into a growing stand. It is used especially for annual crops grown under stands of perennial crops.

Interculture. Arable crops grown below perennial crops.

Mixed cropping. Two or more crops are grown simultaneously in the same field at the same time, but not in row arrangement. (Sometimes called mixed intercropping.)

Relay cropping (relay planting). The maturing annual crop is interplanted with seedlings or seeds of the following crop. If the flowering period of the first crop overlaps with the second crop in the field, the combination becomes intercropping. (Synonymous with relay intercropping.)

Simultaneous polyculture. The simultaneous growth of two or more useful plants on the same plot. This includes mixed cropping, intercropping, interculture, interplanting, and relay planting.

**Terms relating to classification
of the physical environment**

Determinants. Physical or economic variables which determine the performance of cropping patterns.

Environmental complex. A union of locations which share the same values for those physical cropping pattern determinants that have been identified. Synonymous with "agro-ecological analogues."

Agronomic production complex. A union of sites described by values of agronomic determinants in which the relative performance of cropping patterns is substantially similar.

Lowland. Land that is flooded during a major portion of the year when crops are in the field (used especially for flooded rice).

Upland. Land that is cultivated without standing water and which does not flood during the crop season.

Rainfed. Land that is not irrigated.

Field. The largest contiguous unit of land not subdivided by physical or economic restrictions to cultivation of crops.

Socioeconomic terms

Parcel. The largest contiguous unit of land with a given tenure arrangement.

Growth stage. The status of a farming system (on a given farm) with respect to the degree of participation in a market economy, the use of consumer goods, the use of cash inputs, and the degree of farm mechanization.

Resources. The physical (land, light, water, and time, within a climatic environment) and economic (labor, cash, power, and markets) production factors available on a given farm.

Stability. The predictability of a given event or result.

Risk. The lack of stability coupled with the consequences to the farmer of unpredicted poor performance.

Appendix C:
Botanical Names
of Crops Mentioned

banana	*Musa* spp.
black bean	*Phaseolus vulgaris*
breadfruit	*Artocarpus altilis*
cacao	*Theobroma cacao*
cantaloupe	*Cucumis melo*
cassava	*Manihot esculenta*
celery	*Apium graveolens*
chive	*Allium schoenoprasum*
coconut	*Cocos nucifera*
coffee	*Coffea* spp.
cowpea	*Vigna unguiculata*
cucumber	*Cucumis sativus*
ginger	*Zingiber officinale*
groundnut	*Arachis hypogaea*
jackfruit	*Artocarpus heterophyllus*
jute	*Corchorus* spp.
kapok	*Ceiba pentandra*
kenaf	*Hibiscus cannabinus*
lichee	*Litchi chinensis*
maize	*Zea mays*
mango	*Mangifera indica*
millet	*Eleusine indica*
mungbean	*Phaseolus aureus*
papaya	*Carica papaya*
pigeon pea	*Cajanus cajan*
pineapple	*Anasas comosus*
radish	*Raphanus sativus*
rambutan	*Nephelium lappaceum*
rice	*Oryza sativa*
sorghum	*Sorghum bicolor*

soybean	*Glycine max*
squash	*Cucurbita* spp.
sugarcane	*Saccharum officinalis*
sweet potato	*Impomoea batatas*
taro	*Colocasia antiquorum*
watermelon	*Citrullus lanatas*
winged bean	*Psophocarpus tetragonolobus*

Annotated Bibliography

Annotated Bibliography

Chapter 1: A New Approach to Analysis

Rockefeller Foundation. 1976. *Strategies for agricultural education in developing countries.* Agricultural Education Conference II. Working papers. New York.
   A realistic overview of the deficiencies of our present agricultural system which rewards an elite with benefits that do not "trickle down."

Schumacher, E. F. 1973. *Small is beautiful.* New York: Harper and Row.
   This widely read book presents the most forceful social and economic arguments for small enterprise (and small farm) development. The author claims that there is no alternative to "thinking small" and that our development efforts must be reoriented toward "intermediate technology." The book lacks specific suggestions for developing small enterprises.

Wortman, S. 1975. *The world food situation: a new initiative.* A report prepared for the Subcommittees on Science, Research, and Technology and Domestic and International Scientific Planning and Analysis of the U.S. House of Representatives, September 23, 1975. New York: Rockefeller Foundation.
   An outline of development objectives based on "increased production of food and increased purchasing power of masses of the poor."

Chapter 2: The Stages of Small Farm Development

Carneiro, R. L. 1974. The transition from hunting to horticulture in the Amazon basin. In *Man in adaptation: the cultural present,* ed. Y. A. Cohen, pp. 157-166. Chicago: Aldine.
   A case study and discussion of labor productivity as affected by the transition.

De Schlippe, P. 1956. *Shifting cultivation in Africa.* London: Routledge and Kegan Paul.
   An outline of a comparatively simple synthesis of the myriad elements in traditional systems of agriculture. This book provides perhaps the most lucid insight available into the philosophy of traditional agriculture.

Wharton, C. R., Jr. 1969. Subsistence agriculture: concepts and scope. In *Subsistence agriculture and economic development,* ed. C. R. Wharton, Jr., pp. 12-20. Chicago: Aldine.
   An excellent description of the meaning and scope of subsistence farming, which covers 40 percent of the cultivated land of the world and supports 50 to 60 percent of mankind.

Chapter 3: Goals of Small Farm Development

Mellor, J. W. 1969. The subsistence farmer in traditional economics. In *Subsistence agriculture and economic development,* ed. C. R. Wharton, Jr., pp. 209-227. Chicago: Aldine.
   A discussion of farmers' values and objectives and their divergence from national goals.

Schultz, J. L. 1974. Primitive and peasant economies. In *Small farm agricultural development problems,* ed. H. H. Biggs and R. L. Tinnermeier, pp. 61-67. Fort Collins: Colorado State University.
   A discussion of factors motivating small farmers.

Wortman, S. 1974. National agricultural systems. In *Strategies for agricultural education in developing countries.* Agricultural Education Conference II. Working Papers, pp. 20-41. New York: Rockefeller Foundation.
   Includes a discussion of what national development goals should encompass.

Chapter 5: Research in Small Farm Development

Castillo, G. 1969. A critical view of subculture peasantry. In *Subsistence agriculture and economic development,* ed. C. R. Wharton, Jr., pp. 136-142. Chicago: Aldine.
    A discussion of the weaknesses of present research with its one-way flow of information from experiment station to farmer.

Cummings, R. W., Jr. 1976. *Food crops in the low-income countries: the state of present and expected agricultural research and technology.* New York: Rockefeller Foundation.

Gomez, A. A. 1977. Cropping systems approach to production program. In *Proceedings, symposium on cropping systems research and development for the Asian rice farmer,* pp. 441-447. Los Baños, Philippines: International Rice Research Institute.

Harwood, R. R. 1974. Resource utilization approach to cropping systems improvement. In *International workshop on farming systems,* pp. 249-260. Hyderabad, India: International Crops Research Institute for the Semi-Arid Tropics.
    A description of farmer-participant research.

International Rice Research Institute. 1977. *Proceedings, symposium on cropping systems research and development for the Asian rice farmer.* Los Baños, Philippines.
    Summarizes the farmer-oriented research programs of several cropping systems in Asia.

National Research Council. World Food and Nutrition Study Steering Committee. 1977. *Supporting papers: World food and nutrition study,* Vol. 11, Profile 1. A methodology for farming systems research. Washington, D.C.: National Academy of Sciences.
    This report summarizes the inadequacy of conventional, experiment station-oriented research for improvement of farming systems. A transect-analysis method is recommended.

Rockefeller Foundation. 1976. *The role of social sciences in rural development.* New York.

Silvestre, P. 1969. Role of agricultural research in the development of an intensive farming system for Senegal. In *Agricultural research priorities for economic development in Africa*, Vol. 3, pp. 233-239. Washington, D.C.: National Academy of Sciences.

Describes proposed stages of development leading to full integration of livestock and crops.

Chapter 6: Physical Limits to Cropping Intensity

Coulter, J. K.; Derting, J. F.; Oldeman, L. R.; Obradovitch, M. M.; and Slattery, T. 1974. *An agro-climatic classification for evaluating cropping systems potential in Southeast Asian rice growing regions.* Los Baños, Philippines: International Rice Research Institute.

An outline of the crop-oriented climate classification system that has been expanded for use in this chapter.

Duckham, A. N. and Masefield, G. B. 1970. *Farming systems of the world.* London: Chatto & Windus.

The first two chapters discuss factors in the physical environment that affect farming systems.

Francis, C. A. 1979. Small farm cropping systems in the tropics. In *Soil and water management and crop productions, an introduction to regional farming systems*, eds. D. W. Thorne and M. D. Thorne. Westport, Connecticut: Avi Publishing.

An overview of cropping systems in Latin America, the factors influencing them, and the changes that are taking place in them.

Papadakis, J. 1965. *Crop ecological survey in West Africa.* Rome: FAO.

One of the first useful climate classifications based on crop potential.

Wilsie, C. P. 1962. *Crop adaptation and distribution.* San Francisco: W. H. Freeman.

A comprehensive review of physical determinants of the major crops.

Chapter 7: Economic Determinants of Crop Type and Cropping Intensity

International Rice Research Institute. 1975. Constraints to increased rice production. In *International Rice Research Institute Annual Report for 1974*, pp. 266-297. Los Baños, Philippines.

Data on economic determinants of increased rice production.

Jhoda, N. S. 1977. Resource base as a determinant of cropping patterns. In *Proceedings, symposium on cropping systems research and development for the Asian rice farmer*, pp. 101-124. Los Baños, Philippines: International Rice Research Institute.

Lagemann, Johannes. 1977. *Traditional African farming systems in Eastern Nigeria*. München: Weltforum Verlag.
    A comparison of the farming systems of three villages with different intensities of land use.

Price, E. C. 1977. Economic criteria for cropping pattern design. In *Proceedings, symposium on cropping systems research and development for the Asian rice farmer*, pp. 167-179. Los Baños, Philippines: International Rice Research Institute.

Chapter 8: Resource Requirements of Multiple Cropping

Asian and Pacific Council. 1974. *Multiple cropping systems in Taiwan*. Taipei.
    An overview of cropping patterns in Taiwan, and a useful reference on high-input, high-return systems for small farmers.

Bradfield, R. 1966. *Toward more and better food for the Filipino people and more income for her farmers*. New York: Agricultural Development Council.
    One of the earliest accounts by the "father of modern day multiple cropping" of the potential for increased cropping intensity in the tropics.

Dalrymple, D. G. 1971. *Survey of multiple cropping in less developed nations*. FEDR-12. Washington: Foreign Economic Development Service, U.S. Department of Agriculture.
    A good overview of intensive commercial cropping in developing nations.

Greenland, D. J. 1975. Bringing the green revolution to the shifting cultivator. *Science* 190:841-844.
    A list of factors necessary to improve and stabilize production for the African farmer, including use of mixed crops of high-yielding varieties.

Greenland, D. J. 1975. *Evolution and development of different types of shifting cultivation*. Regional seminar on shifting agriculture and

soil conservation in Africa. Soils bull. 24:5-13. Rome: FAO.
   A rational account of the scientific basis for traditional practices
   and their relationship to modern agriculture.

Harwood, R. R. 1976. The application of science and technology to
long range solutions: multiple cropping potentials. In *Nutrition and
agricultural development: Significance and potential for the tropics*,
eds. Nevin S. Scrimshaw and Moisés Behár. New York: Plenum.
   An outline of the crop patterns in traditional agriculture and their
   usefulness for the future.

Innis, D. Q. 1976. Traditional versus modern methods of increasing
tropical food production (in India and Jamaica). General economic
geography, vol. 6. In *Proceedings of the Twenty-Third International
Geographical Congress, Moscow.*
   Questions the attempt to replace traditional mixed cropping
   systems with simplified monocultures.

King, K.F.S. 1968. *Agri-silviculture (The Taungya System)* Bull. no.
1, Department of Forestry, University of Ibadan, Nigeria.
   Describes tree crop associations with annual crops in western
   Africa, listing 79 woody species and 42 agricultural crops grown in
   mixtures.

Levins, Richard. 1973. Fundamental and applied research in
agriculture. *Science* 181:523-524.
   Includes a discussion of multi-storied crop mixtures.

Loomis, R. S. 1976. Agricultural systems in food and agriculture.
Scientific American, 235, no. 3, pp. 20, 98-105.
   A discussion of the technologies relevant to resource-limited
   environments in third world countries.

Norman, D. W. 1974. Rationalizing mixed cropping under indige-
nous conditions: the example of northern Nigeria. *Journal of
development studies*, 11, no. 1:3-21.
   An analysis of the merits of mixed cropping in resource-limited
   situations, in an area where 48 percent of farmers used mixed
   cropping because of higher productivity, and 4 percent gave the
   need for security as a main reason.

Okigbo, B. N., and Greenland, D. J. 1976. Intercropping systems in
tropical Africa. In *Multiple cropping*. Special Publication No. 27,

Madison, Wisconsin: American Society of Agronomy.

Ruthenberg, H. 1971. Farming systems in the tropics. In *Systems with perennial crops*, pp. 189-205. Oxford: Clarendon Press.
An excellent survey of mixed tree systems, their characteristics, and their contribution to smallholder agriculture.

Chapter 9: Animals in Mixed Farming Systems

National Academy of Sciences. 1967. The role of animal agriculture in meeting world food needs. In *Proceedings, 15th annual meeting and minutes of the Agricultural Research Institute, National Research Council, October 10-11, 1966.* Washington, D.C.

Rockefeller Foundation. 1975. *The role of animals in the world food situation, a conference.* New York.

Chapter 10: Noncommercial Farm Enterprises

Stoler, A. 1975. *Garden use and household consumption patterns in a Javanese village.* New York: Dept. of Anthropology, Columbia University

Chapter 11: Nutrient Needs of Intensive Cropping Systems

Lal, Rattan. 1973. Soil erosion and shifting agriculture. In *Shifting cultivation and soil conservation in Africa.* Soils bull. 24, pp. 48-71. Rome: FAO.
A good review of soil tillage and management methods for fragile soils of the high-rainfall tropics.

Nye, P. H. 1961. Organic matter and nutrient cycles under moist tropical forest. *Plant and soil* 13:333-346.
An outline of nutrient cycling in tropical forests, presenting the basic theory of nutrient management suggested in this book for nutrient-poor soils.

Nye, P. H., and Greenland, D. J. 1960. *The soils under shifting cultivation.* Technical communication 51. Reading: Commonwealth Agricultural Bureaux.
A landmark article on the relationships between traditional

farming systems and soil properties, including data on nutrient accumulation in the plant biomass.

Sanchez, P. A. 1977. *Properties and management of soils in the tropics.* New York: Wiley.
 The book has relevance to tropical soils wherever they are found. A chapter on soil management and multiple cropping has special relevance to farming systems.

Sanchez, P. A., and Buol, S. W. 1975. Soils of the tropics and the world food crisis. *Science* 188:598-603.
 A farsighted view of possibilities for dealing with low-nutrient conditions using a minimum of inputs. A significant contribution to thinking about resource-poor agriculture.

Spain, J. M.; Francis, G. A.; Howeler, R. H.; and Calvo, F. 1975. Differential species and varietal tolerance to soil acidity in tropical crops and pastures. In *Soil management in tropical America*, pp. 308-329. Raleigh: North Carolina State University.
 A discussion of crops for low-base-status soils.

Chapter 12: Efficient Use of Farm Resources

Geertz, Clifford. 1963. Two types of ecosystems. In Geertz, Clifford, *Agricultural involution: the process of ecological change in Indonesia*, pp. 13-37. Berkeley: University of California Press.
 A very basic but well-balanced discussion of the paddy and swidden systems. The rest of the book is not recommended.

Rappaport, R. A. 1971. The flow of energy in an agricultural society. *Scientific American* 224, pp. 117-132.
 Case study of energy flow in a subsistence system in New Guinea. Presents arguments on dangers of high energy use, comparing modern with traditional systems.

Wilken, G. C. 1974. Some aspects of resource management by traditional farmers. In *Small farm agricultural development problems*, ed. H. H. Biggs and R. L. Tinnermeier, pp. 47-59. Fort Collins: Colorado State University.
 A discussion of the use of physical resources such as land, water, soil, climate, slope, and space; with comments on the high efficiencies of some traditional systems.

Chapter 13: Requirements for Farm Mechanization

Kline, C. K.; Green, D.A.G.; Donahue, R. L.; and Stout, B. A. 1969. *Agricultural mechanization in equatorial Africa.* Research report 6. East Lansing: Institute of International Agriculture, Michigan State University.

Stresses the differences between power and mechanization. Discusses the social changes required for mechanization of traditional systems. Contains a comprehensive literature review. The best available discussion of the topic.

Chapter 14: Stability of Alternative Systems

Bergeret, Anne. 1977. Ecologically viable systems of production. *Ecodevelopment news* 3. Paris: International Research Center on Environment and Development (CIRED).

An overview of the ecological stability of alternative crop patterns.

Schluter, M.G.G. 1974. *Management objectives of the peasant farmer: an analysis of risk aversion in the choice of cropping pattern, Surat District, India.* Ithaca: Cornell University, Department of Agricultural Economics occasional paper 78.

Describes the avoidance of profit maximization because of risk. Recommends concentrated efforts to increase the ability of small farmers to bear risk.

Wharton, C. R., Jr. 1968. Risk, uncertainty and the subsistence farmer. In *Technological innovation and resistance to change in the context of survival.* New York: Agricultural Development Council.

Describes small farmers as utility maximizers.

Miscellaneous

Mila Medina Ramos. 1976. *International bibliography on cropping systems, 1973-1974.* Manila: International Rice Research Institute.

Sistemas de agricultural tropical. 1974. Turrialba, Costa Rica: Instituto Interamericano de Ciencias Agricolas.

Dayao, B. M., ed. 1977. *Small farm development: a preliminary annotated bibliography of South and Southeast Asian literature covering the period 1970-1976.* College, Laguna, Philippines: Southeast Asian Regional Center for Graduate Study and Research in Agriculture.

Index

Analysis, of farming systems, 6-7
Animals, 93-100
  feed for, 94-97
  management of, 97-99
  as power source, 14, 68, 116-117, 123
  stability of, 126
Annuals
  crop sequencing of, 76-83
  intercropping of, 85-90
  relay planting of, 83-85
  stability of, 126

Banana trees, 102, 104
Batangas, Philippines
  animals in, 96
  cropping system for, 50-52
Biological stability, in farming systems, 126-127
Borer, European maize, 82, 89
Breadfruit trees, 102
Breeding, for insect and disease resistance, 126

Cacao, 90
Carabao, as draft animals, 96
Cash flow
  as indicator of well-being, 29

as limiting factor, 69-70
Cassava
  intercropping of, 86, 88, 102
  planted by subsistence farmers, 12
  as relay crop, 83, 84
Cation-exchange capacity, of soil, 60
Central Luzon, Philippines, 54
Chainaut Research Station, 79-80
Chemicals
  fertilizer. *See* Fertilizers
  for weed control. *See* Herbicides
Chickens, 97
China, research in, 34
Classification system
  environmental, 60-61
  of limiting factors, 45-47
Coconut trees, 13, 90, 92, 102
Coffee, 14, 102
Commercialization, 5
Compost, as fertilizer, 107-109
Consumer stage, of development, 12-15
Cotton, as rotation crop, 82
Cowpea, as rotation crop, 49-50
Credit availability, as limiting factor, 15

Cropping, multiple. *See* Multiple
    cropping
Cropping intensity
    economic limits to, 63-75
    physical limits to, 45-62
Crop residue
    as fertilizer, 107-113
    multiple cropping and, 81
Crop rotation. *See* Rotations
Crops
    choice of: economic determi-
        nants for, 63-75; physical de-
        terminants for, 45-62
    growing period of, 81
    labor profile of, 77
Crop sequencing, 76-83
    disease problems in, 82-83
    insect problems in, 82
    irrigation in, 80-81
    management of, 77
    nutrients in, 79-80
    tillage in, 79-80
    weed control in, 81-82
Cultivation
    mechanized, 122-123
    shifting, 12, 18, 23
Cultural aspirations, effect on de-
    velopment, 25-26

Development
    definition of, 5-6
    goals of, 21-26
    research into, 32-41
    stages of, 9-20
Development programs, 6-8
    traditional, 3-5
Devotions, to deity, as cultural
    aspiration, 25
Diet, as indicator of well-being, 30,
    71
Disease, 82-83, 89
    in fencerow plantings, 104-105
    resistance to, 126
Diversification
    importance of, 75, 117-118
    stability of, 126-127

by subsistence farmers, 12
Downy mildew, 82, 89
Draft animals. *See* Animals
Ducks, 97

Education, children's, as cultural
    aspiration, 25
Energy costs, as limiting factor, 16
Environmental factors, as affecting
    development, 7

Farmer/scientist cooperation, 7, 34-
    37, 41, 130
Farming systems, analysis of, 6-7
Farms
    average size of, 3, 9: as limiting
        factor, 71-73
    layout of, 74-75
    research conducted on, 38-41
Farmyard, use of, 91, 101-103
Feed, for animals, 94-97
Fencerows, plantings in, 103-105
Fertilizers, 69-70, 115, 116
    commercial, 106-107
    compost, 107-109
    mulching, 113
    in multiple cropping, 79-80, 84
Firewood, plantings for, 104, 105
Fruit trees, 90, 102

Geese, 97
Genetic research, 126
Ginger, 102
Gliricidia, 102, 104, 113
Goals, of development, 21-26
    cultural aspirations affecting, 25-
        26
    effect of profit motive on, 22
    labor efficiency as, 23
    long-term versus short-term, 23
    stability as, 22
    sufficient food as, 21
Grain, as chicken feed, 97
Grain dryers, 50
Grazing, of livestock, 95
Groundnut

intercropping of, 86, 88, 89
as rotation crop, 49-50
theft of, 74

Herbicides, 67, 79, 82, 115, 123
Hunting-gathering stage, of development, 11

Imperata cylindrica, control of, 89
Income, investing of, 14, 115
India, hunting-gathering in, 11
Indonesia
animals in, 99
"zero" tillage in, 80
Insects
in fencerow plantings, 104-105
in multiple cropping, 82, 89
resistance to, 126
Insurance crops. *See* Stability
Interactions, in farming systems, 6
Intercropping, 85-90
biological stability of, 127
crops for, 86
insects and diseases in, 89
nutrients needed in, 89
objectives of, 87-89; saving labor, 67-68, 88
weed control in, 89
International Rice Research Institute (IRRI), 34-36, 47, 78, 100
Investing, of income, criteria for, 14-15, 115
Irrigation, 80-81

Jackfruit trees, 102
Japan, research in, 34
Java
fertilizer use in, 116
planting system in, 102

Kapok trees, 102, 104

Labor
displaced by mechanization, 124
efficient use of, 22-23
Labor productivity, 6

as development indicator, 16-17
intercropping and, 88
Land, as limiting factor, 3, 71-73
Land tenure, as limiting factor, 74
Legumes
as animal feed, 96
diseases of, 83
planting of, 80
as rotation crop, 54, 69, 79
trees, 102, 104, 113
Leuceana, 102, 113
Lichee trees, 102
Light intensity, effect on crop productivity, 60
Limiting factors, 45-62
cash as, 69-70
credit as, 15
energy costs as, 16
labor availability as, 64-66, 72
land as, 3, 71-73
management capability as, 66-67, 73
market availability as, 70-71
power as, 15, 67-69
soil fertility as, 17, 59-60
temperature as, 58
tillage capability as, 58-69
water as, 15, 46-48
Livestock. *See* Animals

Maize
as animal feed, 95, 96
cultivation requirements of, 67
diseases of, 82, 89
intercropping of, 86, 87-88, 89
labor requirements of, 64
nutrient requirements of, 69
as relay crop, 83, 84
as rotation crop, 49-50, 80
sweet, theft of, 74
threshing of, 122
Management
of animals, 97-99
in crop sequencing, 77
as limiting factor, 66-67, 117-118
stability of, 127-129

Mango trees, 102
Market availability, as limiting factor, 70-71
Market crops, 14
  theft of, 74
Mechanization, 119-124
  as development indicator, 15-16
  displacing farm labor, 124
  primary, 119-121
  secondary, 121-122
  for transportation, 123
Millet, intercropping of, 86, 88
Mindanao, Philippines, animals in, 99
Mindoro Island, crop diversification on, 12
Moisture. *See* Water
Mulching, to provide nutrients, 113
Multiple cropping, 76-92
  biological stability of, 127
  intercropping, 85-90
  of perennials, 90-92
  relay planting, 83-85
  sequencing, 76-83
Mung bean
  intercropping of, 86, 89
  labor requirements of, 64
  rainfall and, 61-62
  as relay crop, 84
  as rotation crop, 49-50, 80

Nematodes, 82
Nepal
  animals in, 99
  development in, 17-18
  hunting-gathering in, 11
Nitrogen
  in flooded fields, 79
  in leguminous trees, 102, 104, 113
Noncommercial enterprises, 101-105
Nutrients. *See also* Fertilizers
  cycling in crop system, 110-112
  in intensive cropping, 106-114
  in multiple cropping, 79-80, 89

recycling by plants, 109-110
  trees and, 91
Nutsedge, control of, 89
Nut trees, 90

Paddy, elevation of, 56-58
Papaya trees, 102, 104
Pathogens. *See* Disease
People's Republic of China. *See* China
Perennials, multiple cropping of, 90-92
Phosphorus, on flooded fields, 79
Pigeon pea, intercropping of, 86, 88
Pineapple, 102
Plant breeding, 126
Planting
  multiple cropping and, 80
  relay. *See* Relay planting
  water and, 48-49
Power
  animals for, 14, 68, 116-117, 123
  choosing source, 116-117, 119-121
  efficient use of, 123-124
  as limiting factor, 15, 67-69
  mechanical, 15-16, 119-124
Primary mechanization stage, of development, 15-16
Production stability, of farm system, 129
Productivity, intercropping and, 87-88
Profit motive, affecting goals, 22

Rabbits, 97
Radish, as relay crop, 84
Rainfall. *See* Water
Rambutan trees, 102
Recycling, of farm materials, 107-109
Relay planting, 83-85
  crops for, 83
  fertilizers in, 84
  weed control in, 83

Research
  focus of, 32-34
  genetic, 126
  on-farm, 38, 41
  scientist/farmer cooperation, 7,
    34-37, 41, 130
Resistance, to insects and disease,
  126
Resources, efficient use of, 115-118
Rhade tribe, as hunter-gatherers, 11
Rice
  intercropping of, 86, 87-88
  planted by subsistence farmers,
    12
  as relay crop, 83
  rotation crops for, 49-50, 54-56
  sequence cropping of, 78
  threshing of, 122
Rice, paddy (lowland)
  nutrient requirements of, 69-70,
    79
  water requirements of, 48, 53-54
Rice, upland
  intercropping of, 87-88
  water requirements of, 48
Risk. *See* Stability
Rotations, crops for, 49-50, 54-56
Rubber, as market crop, 17

Scientist/farmer cooperation, 7, 34-
  37, 41, 130
Scientists, teamwork among, 7-8,
  62
Security, from theft, 74
Seedbed preparation, importance
  of, 80, 118
Sequencing, of crops. *See* Crop
  Sequencing
Shifting cultivation, 12, 18
  soil fertility and, 23
Siargao Island, development on, 15
Slash-and-burn agriculture, 110
Soil
  drainage of, 49
  types of, 81
Soil fertility

exploitation of, 23
  as limiting factor, 17, 59-60
Sorghum
  as animal feed, 96
  intercropping of, 86, 88
  nutrient requirements of, 69
  as relay crop, 84
  as rotation crop, 49-50, 80
  threshing of, 122
Soybean
  intercropping of, 86
  planting of, 80
  as relay crop, 84
  as rotation crop, 49-50
Stability, 125-130
  biological, 126-127
  economic, 129-130
  as farmer goal, 22, 73-74
  in intercropping, 89-90
  management, 127-129
  production, 129
Stages, of development, 9-20
  early consumer, 12-15
  hunting-gathering, 11
  primary mechanization, 15-16
  subsistence farming, 11-12
Subsistence farming
  as limiting factor, 70-71
  as stage of development, 11-12
Sugarcane, intercropping of, 86, 88
Sweet potato
  as animal feed, 96
  intercropping of, 86
  as relay crop, 83, 84
  as rotation crop, 49-50

Taiwan
  relay cropping in, 83
  "zero" tillage in, 80
Taro, 102
  as relay crop, 84
Tasaday tribe, as hunter-gatherers,
  11
Temperature, as limiting factor, 58
Thailand, animals in, 97, 99
Theft, of market crops, 74

Threshing, 121-122
  contract, 122
Tillage
  importance of, 118
  as limiting factor, 58-59
  power, 67-69, 119-121
  primary, 67, 119-121
  in sequence cropping, 79, 80
  water and, 48
  "zero," 80, 83, 84
Tobacco, 14
Topography, water and, 56-58
Tractor, as power source, 116
  size of, 120-121
Transportation, mechanized, 123
Tree crops, 13-14, 90-91, 101-105
  biological stability of, 126

Vegetables, 69
  labor requirements of, 64-65
  management requirements of, 66

Water
  harvesting and, 49

as limiting factor, 15, 46-58
  planting and, 48-49
  pumping of, 121
  tillage and, 48
  topography and, 56-58
Watermelon
  planting of, 80
  theft of, 74
Weather. *See* Temperature; Water
Weeds
  as animal feed, 96
  control of, 67, 128-129; cultivation for, 122-123; in multiple cropping, 81-82, 83, 89
Well-being
  as condition for investing, 14-15, 115
  indicators of, 29-30
  measurement of, 27-31
Winged bean, 102

"Zero"-tillage cropping, 80, 83, 84